Synthese Library

Studies in Epistemology, Logic, Methodology, and Philosophy of Science

Volume 390

The aim of *Synthese Library* is to provide a forum for the best current work in the methodology and philosophy of science and in epistemology. A wide variety of different approaches have traditionally been represented in the Library, and every effort is made to maintain this variety, not for its own sake, but because we believe that there are many fruitful and illuminating approaches to the philosophy of science and related disciplines.

Special attention is paid to methodological studies which illustrate the interplay of empirical and philosophical viewpoints and to contributions to the formal (logical, set-theoretical, mathematical, information-theoretical, decision-theoretical, etc.) methodology of empirical sciences. Likewise, the applications of logical methods to epistemology as well as philosophically and methodologically relevant studies in logic are strongly encouraged. The emphasis on logic will be tempered by interest in the psychological, historical, and sociological aspects of science.

Besides monographs *Synthese Library* publishes thematically unified anthologies and edited volumes with a well-defined topical focus inside the aim and scope of the book series. The contributions in the volumes are expected to be focused and structurally organized in accordance with the central theme(s), and should be tied together by an extensive editorial introduction or set of introductions if the volume is divided into parts. An extensive bibliography and index are mandatory.

More information about this series at http://www.springer.com/series/6607

Paul Needham

Macroscopic Metaphysics

Middle-Sized Objects and Longish Processes

Paul Needham
Department of Philosophy
University of Stockholm
Stockholm, Sweden

Synthese Library
ISBN 978-3-319-89026-5 ISBN 978-3-319-70999-4 (eBook)
https://doi.org/10.1007/978-3-319-70999-4

Softcover re-print of the Hardcover 1st edition 2017

Printed on acid-free paper

This Springer imprint is published by Springer Nature
The registered company is Springer International Publishing AG
The registered company address is: Gewerbestrasse 11, 6330 Cham, Switzerland

Preface

This book is about matter. It involves our ordinary concept of matter in so far as this deals with enduring continuants that stand in contrast to the occurrents or processes in which they are involved and concerns the macroscopic realm of middle-sized objects of the kind familiar to us on the surface of the earth and their participation in medium-term processes. But the emphasis will be on what science rather than philosophical intuition tells us about the world and on chemistry rather than the physics that is more usually encountered in philosophical discussions. Accordingly, the finer relativistic niceties that come into play at speeds approaching the speed of light, conundrums raised by the undermining of the distinction between intensive and extensive thermodynamic properties in the strong gravitational fields in distant corners of the universe, or the paradoxical nature of the quantum domain won't be troubling us. In so far as science bears on the discussion, it is the everyday science of macroscopic objects characterised by differences of chemical substance and geological and biological complexity rather than the exotic science that is the principal concern of philosophers of physics.

The discussion might be characterised as descriptive metaphysics, being content, to paraphrase Strawson's use of the term, to describe the structure of our scientific thought about the actual world (whereas he said it is "content to describe the actual structure of our thought about the world" (Strawson 1959, p. 9)). The method in the central chapters dealing with the nature of matter will be to pursue key steps in the historical development of scientific thought about chemical substance and related concepts with a view to tracing the emergence of a systematic ontological interpretation. Aristotle's and the Stoics' discussions of elements and mixtures, based on a continuous view of matter, bear some resemblance to modern, macroscopic conceptions and therefore have some interest as precursors to modern views and are of conceptual interest in their own right. Some of the issues they raised are still reflected in modern discussions, despite the considerable increase in complexity of the subject. Other ideas, such as the intimate connection between what modern science distinguishes as substance and phase, maintained their grip on the understanding of matter until the end of the eighteenth century. These are some of the important aspects of the properties of matter that have been neglected by linguistic concerns that have dominated the recent study of mass terms and which the present study seeks to redress. It will be of interest to see how the bounds of possibility are delimited by the laws governing the appropriate use of concepts refined for the purpose of describing the behaviour of matter. But the concern with modality will not stretch to venturing into the realms of metaphysical possibility governed by philosophical speculation about individual essences and underlying natures.

Like many contemporary discussions of material objects, this one relies heavily on mereology. Unlike several such discussions, this one adheres to the classical principles of mereology governing the usual dyadic relations of part, overlapping, etc. and the operations of sum, product and difference. These principles apply to the mereological structure of regions of space, intervals of time, processes and quantities of matter. Deviations from classical mereology are often motivated by appeal to circumstances applying specifically to material objects understood to gain and lose parts, which some authors take to call for a modification of the classical dyadic mereological relations to triadic relations with the introduction of a third term referring to time and others see as challenging the extensionality of mereology. Although the proper treatment of the temporal aspect of the features of material objects is a central issue in this book, however, the topics falling under this heading are not addressed by relativising mereological concepts to time. Rather, a distinction is drawn between quantities of matter, which don't gain or lose parts over time, and individuals, which are typically constituted of different quantities of matter at different times. This distinction undermines the motivation for challenging what is usually understood as the extensionality of mereology.

These general indications of scope and limitations of the study are now supplemented with a brief synopsis of the various chapters, giving an overview of the main lines of argument and providing some hints about details some readers might like to skip, at least on a first reading.

Synopsis

An ontology of continuants and occurrents to be developed in the course of the book within a framework of regions of space and intervals of time is initially outlined in Chap. 1. Much of the chapter is concerned with explaining how mereology is understood. First and foremost, quantities of matter, to which the principles of classical mereology are held to apply, are distinguished from material objects, here called individuals, that change their constitutive matter over time and to which mereological principles don't apply. The motivation for this and the detailed development of the features of the constitutes relation (as distinct from mereological parthood) is discussed in Chaps. 2 and 3. On this understanding, certain lines of objection to classical principles of mereology are put aside. Rehearsing reasons for rejecting some other suggestions for modifying classical principles of mereology serves further to illustrate how the mereological concepts are understood here. But the main thrust of the chapter is to emphasise what the mereological principles say and what they leave open concerning the relations of part, overlap, separation and identity and the operations of sum, product and difference. Classical mereology is an incomplete theory whose axioms can be supplemented in various ways to characterise the kind of objects to which they apply. It is shown how this can be done for a theory reasonably called a theory of temporal intervals and again when supplemented with an additional nonmereological primitive to develop a theory of spatial regions. Both theories are complete first-order theories (i.e. no further independent axioms can be added without contradiction). An analogous development doesn't seem possible for a pure mereological theory of quantities of matter. Developing a theory of quantities of matter requires introducing times and spaces, as well as other entities, along with predicates relating these various kinds of entities, which is pursued in the following chapters. Readers who are not interested in the technical details may like to skim quickly over the presentations of the theories of times and spaces, together with the proof of completeness, in Sects. 1.3 and 1.4 and their corresponding subsections, and proceed directly to the final section of the chapter which outlines the strategy for the remainder of the book.

A fundamental relation connecting matter with space and time is the occupies relation. The occupies relation stands between a material body, a region of space and an interval of time, and its features and related concepts are pursued in Chap. 2. Individuals are the material bodies considered first, and it is argued that it is unnecessary to consider the regions they occupy to be bounded by boundary entities. An account of the abutment of individuals is presented on the basis of the spatial regions described in Chap. 1 without needing to introduce boundaries as distinct entities. This is a time-dependent relation which treats times as intervals and accommodates the fact that the bodies in question might move whilst abutting, calling on an account of the occupies relation that accommodates movement. What I call an accumulation condition, formulated in mereological terms, is introduced for this purpose. Analogues of this mereological condition for other predicates are discussed in the sequel.

It transpires that individuals don't in general occupy the same regions as the quantities of matter that constitute them. Although a head might be said to abut a torso (Smith and Varzi's example) or the atmosphere abuts the sea (Leonardo), this doesn't mean that the matter constituting these bodies abuts during the time at issue. Blood circulates between head and torso, and matter (water, CO_2, etc.) is exchanged between the sea and the atmosphere. This calls for a distinction between individuals (the sea, the atmosphere) and the corresponding quantities of matter of which they are constituted (seawater, air).

The constitution relation is the subject of Chap. 3. It is a three-place relation standing between an individual, a quantity of matter and a time, allowing for the typically changing constitution of individuals over time. The fact that times are intervals is accommodated and the relations between the occupies relation are explored, in terms of which coincidence is defined. Time-dependent analogues of the mereological relations are defined as triadic relations between two individuals and a time, but there are no corresponding analogues of the mereological operations. Modal features of individuals and quantities are compared. An appendix criticising van Inwagen's understanding of parthood as time dependent concludes the chapter.

The following four chapters take up the features of quantities of matter, distinguished from individuals by their mereological structure. This structure is reflected in the character of certain predicates expressing the property of being a particular kind of substance, say, water, and exhibiting one or more phases, such as being liquid and being gas. Such predicates seem to ascribe an amorphous character to their subjects, which authors have tried to more precisely capture in terms of a distributive and a cumulative condition, formulated in terms of the mereological part relation and sum operation, respectively. These conditions are presented in Chap. 4 and generalised to apply to relational predicates. This generalisation is obviously needed for relational predicates like "is the same substance as" and "is warmer than". But substance and phase predicates, by contrast with the monadic predicates "is a time" and "is a spatial region", are also taken to be relational. The relational interpretation follows the chemist's understanding of the permanence of matter throughout chemical change, which has been standard at least since the time of Lavoisier. Aristotle and the Stoics can be reasonably interpreted, it seems, to have thought that the distributive condition holds for substance and phase predicates, as discussed in Chap. 5. Whether the conditions can be upheld from a modern perspective is discussed in Chap. 7.

Substance and phase predicates fall under the linguistic category of mass predicates in accordance with grammatical criteria for their use with articles in singular inflexion and qualification as much/little rather than many/few. This category is more extensive and includes terms whose normal translations into some foreign languages don't comply with the grammatical criteria for mass predicates whilst lacking terms whose normal translations into some foreign languages do comply with the grammatical criteria. It is therefore fallacious to argue that features not applying to all mass predicates (of a particular natural language) don't apply to substance or phase predicates. But whilst linguistically inspired characterisations of mass predicates don't necessarily say anything about the nature of substances and phases, Quine presents a general reason for rejecting the distributive condition on all mass predicates which he specifically illustrates with substance predicates based on his understanding of what science tells us about the atomic nature of matter. Quine's case is discussed in Chap. 4, but as we will see in Chap. 7, it isn't the last word on the import of the microscopic view of matter presented by modern science for the distributive condition.

Taking advantage of what has been said about the occupies relation in Chap. 2, versions of the general distributive and cumulative conditions restricted to spatial parts are formulated. This raises the issue of co-occupancy, and the final section of Chap. 4 reviews some of what has been about this and the very little argument against the feasibility of the idea.

Ancient theories of substance based on a continuous view of matter are taken up in Chap. 5. They bear some resemblance to modern, macroscopic conceptions and therefore have some interest as precursors of modern views and are of conceptual interest in their own right. Aristotle's views of the matter were based on the fundamental tenet that a homogeneous quantity comprises a single substance. Combined with the thesis of the impossibility of co-occupancy (disputed by the

Stoics), he was thus able to provide a criterion for a quantity comprising a single substance (i.e. being a pure sample of a substance), namely, one whose all parts are of the same substance. He distinguished four elements, characterised by properties they exhibit in isolation, which combine in one of two kinds of mixing process that he distinguished to yield other substances. It follows from his fundamental tenet that these derived substances have no parts that are elements. The elements are held to be potentially present in the sense that a suitable "decomposition" process would change the quantity into one that is partly one element, partly another and so on for the four elements. There are suggestions and problems associated with this account that are not developed or resolved in the extant texts, but some of the central ideas are clear enough and remain at issue in some related guise in modern chemical theory. The Stoics thought the elements are present in their compounds (blends), which agrees, it has been said, with the modern view. This led them to accept the possibility of co-occupancy and claim that the elements occupy the same place as one another and the blend which they form (which doesn't entail that an element occupies the same-sized region—of the same volume—when combined as when isolated). However, they failed, as far as I can see, to address the issue of giving a characterisation of the elements that is not only applicable when in isolation. The chapter closes with a discussion of the distinction between potential parts and potential qualities in connection with Holden's recent book on the early modern discussion of the infinite divisibility of matter.

An important feature of the transition from ancient to modern views of matter is recognition of the distinction between substance and phase and the realisation that a substance like water is not necessarily liquid, as Aristotle thought, but the same substance might be solid or gas too. As discussed in Chap. 6, the older view was surprisingly tenacious, and even Lavoiser had difficulty abandoning it completely although he prepared the ground for doing so. The distinction was clear enough throughout most of the nineteenth century but received its definitive formulation in Gibbs' phase rule which he presented in connection with his formulation of chemical thermodynamics in the late 1870s. The import of this fundamental principle of chemistry is discussed and illustrated in the final sections.

With this background, the dyadic predicate "water" is discussed in some detail in Chap. 7 as illustrative of substance predicates. The familiar claim "Water is H_2O" cannot be construed as a claim about the constitution of water at the microlevel but is a purely macroscopic claim about what chemists call the composition of water. Relating the macroscopic term "water" to microscopic constitution must take cognizance of the fact that the situation is a dynamic one, not a fixed, unchanging structure. The continual restructuring renders macroscopic quantities effectively homogeneous for macroscopic times, say, of the order of a millionth of a second or longer. (Spectroscopists describe the structure accessible over shorter times via higher frequency radiation as indiscernible—blurred—over the longer times probed by lower frequency radiation.) What should we say about extremely small parts of a quantity of water for extremely short subintervals of time?

Infinite divisibility of time entails that intervals have proper subintervals without limit. A uniform treatment of quantities is less obviously appropriate in moving from the macro- to the microdomain of quantum mechanically indiscernible entities. Nevertheless, a uniform treatment is considered, bearing in mind that the parts are parts of a macroscopic quantity for the times at issue and not isolated from the bulk matter. But an alternative treatment is developed, where following French and Krause's two-sorted interpretation in which identity is meaningfully predicated of macro- but not microentities, the mereology of quantities is restricted to exclude microentities as parts. This is consistent with the conclusion that might be drawn from the discussion in Chap. 4, that the quantity, time and space variables to which predicates at issue here are affixed range over macroscopic domains.

Phase predicates don't follow the same pattern, mereologically speaking, as substance predicates. The microscopic understanding of the dynamic equilibrium between the exchange of matter and abutting phases calls for a different mereological interpretation, and an accumulation condition similar to that applying to the occupies relation comes into play.

Processes are introduced in Chap. 8. In the wake of a discussion of the notion of change, any idea of introducing occurrents on the basis of a distinction between intrinsic and Cambridge changes in continuants is abandoned. The understanding of occurrents as temporally extended processes—causings rather than relata of a dyadic causal relation—based on the analysis of thermodynamics in the second half of the nineteenth century is conceptually more illuminating. The chapter continues with a development in the mereological features of processes. Mereological properties include distinguishing temporal parts, parts arising from a consideration of the parts of bodies involved and parts of complex processes such as the elementary reactions underlying complex chemical reactions. The relational character of processes (arising from the several bodies typically involved in processes) and their modal character are taken up in the final sections.

Modal features of quantities are discussed in the final chapter, beginning with a discussion of how resilient the same substance, in particular the same element, is to possible changes in states of combination. The chapter continues by pursuing the analysis of the idea of being possible for a quantity to be such and such. Claims difficult to capture with conventional sentential modal operators lead to the use of possible states to achieve appropriate expressive power. Incorporation of possible processes is a natural development, linking up with the discussion in the previous chapter.

The various citings of Duhem's works throughout the book give some indication of the source of inspiration that he has provided. I would also like to acknowledge and give thanks for the support from several of my contemporaries. Many subjects discussed in this book have been aired at various conferences and seminars. I am grateful to audiences at the meetings of the Baltic Workshop for Logic and Philosophy of Science, the International Society for the Philosophy of Chemistry, the (American) Philosophy of Science Association, the European Philosophy of Science Association and conferences and seminars at the Universities of Bristol, Copenhagen, Durham, Geneva, Ghent, Gothenburg, Leuven, Stockholm and Uppsala. But I would particularly like to thank Robin Hendry, Jaap van Brakel, Stephen Weininger, Michael Weisberg and my colleagues at Stockholm Sören Häggqvist,

Dugald Murdoch and Åsa Wikforss, for lending a sympathetic ear. Much of the material in the book has been discussed in a course I taught at Stockholm University, and I am grateful for the enthusiastic response of students who participated. An anonymous referee provided some much-appreciated comments which were helpful in finalising the manuscript. Finally, thanks to Otavio Bueno for putting me in touch with Chris Menzel, who helped me sort out some of my problems with Latex.

Acknowledgements

Various portions of the book derive from some of my previous publications, although with revisions and additions when forming part of the book. The material I have drawn on comes from the following sources:

- Section 1.3 is based on "Temporal Intervals and Temporal Order", *Logique et Analyse*, 93 (1981), 49–64. Reprinted by permission of the editor.
- Substantial parts of Chap. 3 are based on "Transient Things and Permanent Stuff", *Australasian Journal of Philosophy*, 88 (2010), 147–66. Reprinted by permission of Routledge.
- Parts of Chap. 5 are based on "Macroscopic Mixtures", *Journal of Philosophy*, 104 (2007), 26–52, by permission of the editors.
- Parts of Chap. 5 are based on "Duhem's Theory of Mixture in the Light of the Stoic Challenge to the Aristotelian Conception", *Studies in History and Philosophy of Science*, 33 (2002), 685–708; "An Aristotelian Theory of Chemical Substance", *Logical Analysis and History of Philosophy*, 12 (2009), 149–64; and "Méréologie des substances chimiques et de leur transformation: une approche analytique", forthcoming in *La chimie, cette inconnue?*, ed. Jean-Pierre Llored, Editions Hermann, Paris, 2016. All reprinted by permission of the editors.
- Section 6.4 draws on "A Mereological Interpretation of the Phase Rule", *Philosophy of Science*, (2010), 77, 900–10, by permission of the editors.
- Sections 7.1, 7.2 and 7.3 draw on "Substance and Time", *British Journal for the Philosophy of Science*, 61 (2010), 485–512. Reprinted by permission of Oxford University Press.
- Chapter 8 draws on "Macroscopic Processes", *Philosophy of Science*, 66 (1999), 310–331, and "Process and Change: From a Thermodynamic Perspective", *British Journal for the Philosophy of Science*, 64 (2013), 395–422, by permission of the editors and Oxford University Press.

Stockholm, Sweden
April 2016

Paul Needham

Reference

Strawson, P. F. (1959). *Individuals: An essay in descriptive metaphysics*. London: Methuen.

Contents

1 **Mereology** ... 1
1.1 Continuants and Occurrents ... 1
1.2 Mereology ... 5
1.3 Times ... 11
1.3.1 Directed Time ... 15
1.4 Regions of Space ... 18
1.4.1 Completeness of the Theory of Regions ... 21
1.5 Taking Stock ... 23

2 **Occupying Space** ... 29
2.1 Boundaries ... 29
2.2 Doing Without Boundaries ... 31
2.3 Keeping in Touch ... 32
2.4 Abutment Without Material Abutment ... 35
2.5 Mathematical Analysis and Indispensability ... 38
2.6 Occupying a Region Exactly ... 44

3 **Constitution** ... 47
3.1 Distinguishing Constitution and Parthood ... 47
3.2 Individuals ... 51
3.3 Occupation for Individuals ... 56
3.4 Component Parts ... 59
3.5 Operations on Component Parts? ... 61
3.6 Modality ... 66
3.7 Appendix: Van Inwagen's Understanding of Parthood as Time-Dependent ... 71

4 **Distributivity and Cumulativity** ... 75
4.1 Mass Predicates ... 75
4.2 The Distributive and Cumulative Conditions ... 77
4.3 Relational Predicates ... 78

4.4 Generalising the Cumulative Condition ... 81
4.5 Spatial Parts ... 83
4.6 Is Cooccupancy Really Possible? ... 86

5 The Ancients' Ideas of Substance ... 89
5.1 Doing Without Atoms ... 89
5.2 Aristotle's Conception of Substances as Homogeneous ... 91
5.3 Two Kinds of Mixing Process ... 93
5.4 Simplicity, Actuality and Potentiality ... 96
5.5 The Stoic Alternative to Potential Presence ... 99
5.6 Formalisation ... 101
5.7 Potential Parts and Potential Qualities ... 106

6 The Nature of Matter ... 113
6.1 The Scope of Metaphysics in the Enlightenment ... 113
6.2 Phase-Bound Substances ... 115
6.3 Distinguishing Substance and Phase ... 118
6.4 Thermodynamics and the Phase Rule ... 120

7 The Relation of Macroscopic Description to Microstructure ... 127
7.1 Water is H_2O ... 127
7.2 What Do Substance Predicates Apply To? ... 129
7.3 What Do Phase Predicates Apply To? ... 138
7.4 Indiscernible Particles ... 142
7.4.1 Discussion ... 149

8 Longish Processes ... 151
8.1 Occurrents ... 151
8.2 Change ... 153
8.3 Distinguishing Processes from States ... 158
8.4 Causings ... 164
8.5 Relating Processes to Quantities, Regions and Times ... 166
8.6 The Mereological Structure of Processes ... 169
8.7 The Relational Character of Processes ... 172
8.8 Modality ... 176

9 Modal Properties of Quantities ... 179
9.1 Introduction ... 179
9.2 Elements and Compounds ... 180
9.3 Water is H_2O Again ... 185
9.4 Possible States ... 188
9.5 Possible Processes ... 190

Appendix Summary of Some Main Principles ... 201
Mereological Axioms for Quantities ... 201
Abutment ... 202

Occupancy ... 202
Static Abutment ... 203
Abutment on the Move ... 203
Constitution ... 204
Lavoisier's Principle of the Indestructibility of Matter ... 205
Distributivity and Cumulativity ... 206
Processes ... 207
Modal Properties of Quantities ... 208

Glossary ... 209

Bibliography ... 211

Index ... 219

Chapter 1
Mereology

1.1 Continuants and Occurrents

Middle-sized objects range from things very large to things very small on a human scale, and are included in what science counts as macroscopic objects and treats as classically identifiable. They are the material objects that metaphysics has traditionally been concerned with and what most of the world's scientists are engaged in studying. In the terminology of Johnson (1921) they are continuants, with the general feature that they may have different properties at different times, or stand in different relations to different times. A continuant might be warm at one time and cold at another, subject to a pressure of one atmosphere at one time and to half that at another, and so on. Johnson distinguished continuants from occurrents, which don't persist over time although they extend over time. It is their different temporal parts, and not they themselves, that have different properties at different times. A more familiar term is "processes". Philosophers more usually speak of events. But although no distinction between the two is developed here, the idea of a process continuously unrolling over time is more suggestive of their character than that of the event of an accomplished action with no temporal extension.

Three-dimensionalists, as advocates of continuants in the fundamental ontology are sometimes called, think that continuants have no temporal parts. Rather, they endure from one time to another in the sense that the same (identical) object has one property at the one time and possibly another at the other. This claim is sometimes glossed as saying that continuants are "wholly present" at a time when they exist. The use of identity is hardly elucidated by this expression, which I venture to say offers no improvement on the statement that objects endure in the sense that *they* may have different properties at different times. Only processes (Johnson's occurrents), which don't endure over time and have temporal extension, have temporal parts. Earlier and later parts of a process are proper parts of that process, and not identical with one another nor with the whole process.

P. Needham, *Macroscopic Metaphysics*, Synthese Library 390,
https://doi.org/10.1007/978-3-319-70999-4_1

The autonomy of these two categories has been questioned. Philosophers such as Lombard (1986) and Lowe (2006) have argued that occurrents are but changes in continuants, and so reducible to changes in continuants (reduction thesis 1). Others, such as Donald Davidson, and philosophers standing in the tradition of "process philosophy" after A. N. Whitehead, have argued for the importance of recognising events or processes in the basic ontology, some going so far as to argue for a reduction of continuants to processes (reduction thesis 2) although not Davidson. I'm with Davidson on this point (although not on several other points concerning occurrents), and processes will be systematically introduced here in due course as autonomous entities alongside continuants.

This general scheme of continuants and occurrents has been subject to a more radical challenge from so-called four-dimensionalists, such as Heller (1990), Sider (2001) and Hawley (2002). They argue that the only physical entities are four-dimensional ones which, rather than being warm at one time and cold at another, have a temporal part which is warm and another temporal part which is cold. According to the four-dimensionalist's account of how ordinary objects persist over time, it *perdures* by virtue of really being a four-dimensional object. Some four-dimensionalists (so-called "worm theorists") think that perduring from one time to another is a matter of being different temporal parts of an extended four-dimensional object. Other four-dimensionalists (so-called "stage theorists") have it that what actually exist are instantaneous four-dimensional entities, and perduring from one time to another is a matter of different stages standing in an appropriate counterpart relation. Perduring, whether on the worm-theorists' or the stage-theorists' account, stands in contrast to the three-dimensionalists' conception of persistence over time for which the term "endurance" is used in the literature. If four-dimensional entities are events or processes, as textbook accounts of the four-dimensional Minkowski space-time of special relativity often suggest, then the four-dimensionalist's challenge might be subsumed under reduction thesis 2. Four-dimensionalists adopt very different arguments from the process philosophers, however.

There are many varieties of three-dimensionalism, quite apart from whether their proponents advocate reduction thesis 1 or not. Some don't accept the distinction that will be maintained here between an individual and its constitutive matter,[1] arguing that "constitution is identity", as Noonan (1993) puts it. Even some of those who do put it in a radically different way from the way I will develop it (e.g. Baker 1997, 2000). Many of the disputes hang on the understanding and application of mereology, a formal theory of parthood which will figure heavily in the present investigation. One of the main sticking points is the operation of summation, involving in its most general form the unique existence of the sum of objects satisfying an arbitrary condition. Some philosophers would have a restriction to spatially connected objects; others claim the attempt to formulate a summation

[1]Following Chappell (1973) in the use of the term "individual" for what he distinguishes from "matter".

principle with restrictions is incoherent. Some deny the existence of sums and maintain that all that exists are the basic entities of their ontology.

As this discussion illustrates, authors often enter this arena with very substantial initial commitments, and it would be overly naive to expect that arguments can always, or even usually, be resolved by appeal to easily identified neutral ground. Explicitness is, of course, a prime virtue in this context. I will be particularly concerned with the conceptual development of the scientific study of the matter constituting individuals of variable composition, from the ideas of Aristotle and the Stoics on the nature of substance and mixture to modern chemical theory. Where this conceptual history is not clearly at issue in the preparatory chapters setting up the framework for this discussion, I will be predisposed to take a lead from natural science and avoid metaphysical speculation about essence. The major concern will be with the medium-sized objects and longish processes typified, but by no means exhausted by, the entities we see around us. Science has applied itself to the study of such things, improving the precision of everyday claims by explicitly and ever more carefully specifying the limits of error within which systematically used concepts are applied (cf. Duhem 1954, pp. 174–9, 190–5), and delivered a theoretically well-developed conception of macroscopic phenomena which is empirically supported up to the articulated limits of error.[2] This is not the glamorous science of the scientific revolution at the beginning of the twentieth century. Accordingly, since the present study won't be overly concerned with bodies approaching one another at near-luminal velocities, the corresponding spatio-temporal considerations will not be a priority. Nor is the study primarily concerned to describe the features of bodies in a context where the gravitational forces are so large as to seriously infringe on the notion of additivity on which the thermodynamic distinction between extensive and intensive properties is based. At the other limit of applicability of thermodynamics, it will not be dealing with matter on such a small scale that statistical fluctuations become significant. And it will certainly not be entering the domain of such short times that the time-energy uncertainty principle of quantum mechanics comes into play. A priori speculation about what holds at instants will accordingly be avoided here. The object is to provide an account of the ontology of continuants and occurrents as these figure in middle-range, macroscopic phenomena, recognising that this may not be straightforwardly applied or extended to entities outside this range. Others may have the ambition of embracing a broader ontology. I will only say that, just as a claim isn't scientifically precise simply by virtue of omitting to make a reasonable claim about limits of error, so this aim is not ensured merely by omitting to acknowledge the limits of applicability of relevant empirically based theories.

[2]The reference to Duhem reminds us that concern with the macroscopic realm is not equivalently expressed as a concern with the observable as opposed to the theoretical. Two points about his fundamental argument against a rigid distinction between the two (1954, pp. 144–64) are noteworthy here. His holistic argument applies to the macroscopic domain, and the holism is restricted. It is not the global holism of Quine's "Two Dogmas" which has confused Duhem's point with the so-called "Duhem-Quine" thesis which sanctions the underdetermination thesis.

Medium-sized, moderately long-lived inhabitants of the macroscopic realm, from which philosophers usually draw for their examples, are sometimes said to be eliminable in favour of a more fundamental realm of microscopic entities. But the arguments for reduction on which such views rely no longer receive the broad, unreflecting support that they once enjoyed. The underlying mechanical vision driving such claims is said to rely on extra-mechanical principles whose sole motivation seems to be the anticipated reduction (Sklar 1993, pp. 371–2). Standard reviews of statistical mechanics continue to list the seemingly insurmountable problems facing the would-be reductionist (amongst which is the fact that the standard practice of physicists is to use the Gibbs formulation of statistical mechanics, whereas a reduction would require the Boltzmann approach—see Frigg 2008). Belief in reductionism seems more like an existential leap of faith than an appeal to considered scientific judgement.

Recognition of the microworld is not sufficient to motivate eliminative ontological reductionism. Another view of the relation between micro- and the macroscopic theory is that they work in tandem. I have argued elsewhere that this is how the standard example of reduction, the relation between gas temperature and the average kinetic energy of the gas molecules, should be viewed, whereas construing the relation as reductive is circular (Needham 2009b, 2010). If reductionism is not plausible in this supposedly paradigm case of reduction, it certainly isn't more plausible in more complicated cases. Science cannot even specify the microscopic correlates of the thermodynamic properties of water beyond a rough qualitative description. It is if not tautological then a scientific truism to say that the microstructure of water is arranged "waterwise" because we've known of the existence of the microrealm since Perrin's investigations of Brownian motion. The challenge of reduction is to rise above such platitudes, and nothing short of an adequate demonstration will do.[3]

[3]Philosophers with a special agenda may seek to circumvent the requirement of providing the details of a successful reduction by handwaving generalisation from a simple case that in no way justifies the general eliminative claim. This is not to deny that it may provide motivation for the proponent to pursue the line of enquiry and seek to flesh out the view. But it is in no way adequate to block the pursuit of alternative views not sharing the same vision. Thus, as a step in the presentation of his ontological structuralism, French (2014, Ch. 7) suggests that the general notion of the solidity of material objects is to be understood in terms of, and so eliminated in favour of, the way the Pauli exclusion principle (the antisymmetry requirement) governs the electronic structure of molecules. This leaves untouched the detailed problem of reducing to the properties of elementary particles the melting point of water, the point of maximum density of liquid water, and so on for virtually all the specific properties characterising macroscopic objects. In a later chapter he trades on the difficulties of characterising individual biological organisms (see e.g. Dupré and O'Malley 2009) to indicate how the structuralist viewpoint he advocates might be applicable outside the narrow confines of quantum physics, but without offering any definitive proposal of comparable detail.

In this connection it is interesting to note that the concept of an organism has once more become central to biology, having been denigrated in the twentieth century by population statistics and molecule biochemistry to "nothing more than an epiphenomenon of its genes". Now, organisms are taken to be the primary agents of evolutionary change and the reductionist tenets of the central dogma, with its excessive reliance on molecular-level explanations, have given way to the view that DNA replication, protein synthesis, etc. proceeds "by virtue of the enabling conditions afforded by the pre-existing fundamental organization of the cell as a whole" (Nicholson 2014, pp. 348, 353).

The notion of a part plays a central role both in the conception of those continuants referred to here as quantities of matter and of processes developed in later chapters. It figures differently in the understanding of some of those who take a different view of these matters. For the sake of clarity, then, even if it would be too much to hope to circumvent controversy, the following sections endeavour to explain how mereology, the formal theory of parthood, is understood and applied here and what position is taken on some of the controversies concerning the application of the theory. The chapter concludes with a statement of the strategy for approaching the general features of quantities of matter in the light of the limitations of purely mereological concepts. This is based on a contrast with what can be done by way of elaborating the character of times and spaces. Readers who might like to gloss lightly over the technical details in the mereologically-based theories of times and spaces in Sects. 1.3 and 1.4 will understand the strategical points if they simply accept the completeness of these theories.

1.2 Mereology

The term "mereology" was coined by the Polish logician Lesniewski (1916) on the basis of the Greek $\mu\epsilon\rho o\varsigma$, "part". The term was taken to refer to an algebraic theory of parthood which Lesniewski sought to use as a way of bypassing Russell's paradox in the theory of sets (see Simons 1987, p. 102). The theory has since been imported into analytic philosophy as the basis of understanding "concrete" objects, as distinct from "abstract" objects of the kind mathematics is taken to deal with and typically exemplified by sets. It inspired Nelson Goodman's nominalism, which was an attempt to circumvent as far as possible the need to appeal to an ontology of abstract objects.

General ideas captured by the system of mereology appeared before Lesniewski's work. Potter (2004, p. 23) says that "Frege … deserves credit for having first laid out the properties of fusions [sums] clearly. A fusion, he said, 'consists of objects; it is an aggregate, a collective unity, of them; if so it must vanish when these objects vanish. If we burn down all the trees in a wood, we thereby burn down the wood. Thus there can be no empty fusion.' (1895, pp. 436–7, modified)". Potter continues, "it was plainly fusions, not collections, that Dedekind had in mind in *Was sind und was sollen die Zahlen?* when he avoided the empty set and used the same symbol for membership and inclusion …—two tell-tale signs of a mereological conception. He drafted an emendation adopting the collection-theoretic conception only much later".

While set theory went its own way after Zermelo's axiomatisation of the notion of collection, the sum concept was developed in algebraic fashion in, for example, the early paper of Huntington (1904). In what is possibly the first application of mereology, Huntington (1913) develops a formulation of Euclidean geometry on the basis of a notion of a region of space rather than a point, with two primitives "inclusion" and "sphere". Although he didn't use the term "part" it is clear that it is

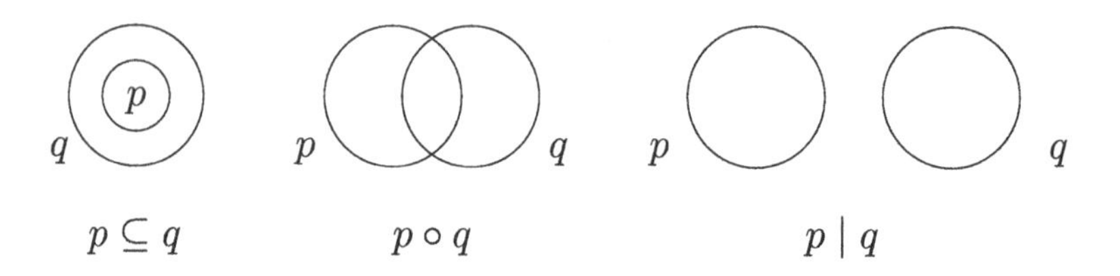

the mereological relation he had in mind, and this was made explicit in Tarski's short 1926 paper designed to accomplish essentially the same task. Regions of space are the paradigm examples of mereologically related entities, and mereological relations such as being a part of, overlapping and being separate from are naturally pictured as relations between regions in the accompanying figure.

The word "mereology" is naturally regarded as a technical term, referring to an algebraic system in this tradition, and whether mereology is applicable to the elucidation of particular kinds of entities depends upon whether the mereological relations (part, overlap, separate from, etc.) and operations (sum, product, difference) are applicable with the features prescribed by a system of this kind. Where these features don't apply, the concepts concerned are not mereological concepts. This is how the term is used here. Some philosophers have in recent times used the term differently, referring to any use of the word "part" or an expression deemed to have some such import as mereological. Noting that ordinary usage doesn't always comply with the traditional mereological principles, these principles are questioned and modifications sometimes suggested. Objections against what is called the mereological principle of extensionality, otherwise known as the mereological criterion of identity (same parts, same thing), motivated by regarding the constitution relation as mereological parthood are dealt with here by giving a detailed treatment of the constitution relation (Chap. 3) on the understanding that it is not a mereological relation. Other targets for criticism have been the mereological principle of the transitivity of the part relation and the existence of unrestricted sums.

Rescher (1955, p. 10) started the trend with the claim that "There are various nontransitive senses of 'part.' In military usage, ... persons can be parts of small units, and small units parts of larger ones; but persons are never parts of large units. ... A part (i.e., a biological sub-unit) of a cell is not said to be a part of the organ of which that cell is a part". Johanna Seibt (2000, p. 256–7) adds other examples: "The moulding is part of a door. The door is part of a house. *The moulding is part of a house." Again, "Changing diapers is part of being a mother. Opening the box with wipes is part of changing diapers. Pressing your thumb upward is part of opening the box with wipes. *Pressing your thumb upward is part of being a mother." David-Hillel Ruben tries to put more flesh on the bones with his example:

> Suppose I have an alarm clock, one of whose parts is a light by which one can tell the time at night. Suppose that I use the clock to make a bomb that detonates when the clock's hands reach a specified position. The light makes no contribution to the working of the bomb. The light is one of the parts of the clock, and the clock is one of the parts of the bomb, and yet it is false that the light is one of the parts of the bomb. This seems to show that the relation of being a part of is non-transitive. (Ruben 1983, p. 231–2).

There are two things to say about examples such as these. First, as putative counterexamples to the transitivity of the part relation they are remarkably unconvincing. Take the last example. If Ruben really means both that the light is part of the clock and the clock is part of the bomb, then I cannot for the life of me see how he can deny that the light is part of the bomb. It seems that in affirming the falsity of the conclusion he is interpreting the one premise, that the clock is part of the bomb, to imply the denial of the other, that the light is part of the clock. There is a simple equivocation on the term "clock" as the contraption that includes the light and as a more specialised device serving as a detonator. The other examples are no more plausible. It is not easy to see that the first premise, "The moulding is part of a door", deals with the same relation as the second, "The door is part of a house", in one of Seibt's putative counterexamples. The first is an idiomatic way of expressing that the door has a certain kind of shape and can hardly be interpreted as literally ascribing the part relation at issue in the second premise to a pair of objects. A clear and unambiguous interpretation of the premises in the other putative counterexample of hers quoted above is also wanting. What sort of thing is being a mother? Is it to be thought of as a universal, or perhaps as a process? A process interpretation, which seems to be at issue in the other premises, is perhaps not the most natural interpretation of the first premise or the conclusion. But uniformly interpreted true premises and a false conclusion is what is required, and it is far from clear that this is such a case.

Second, even if these or other examples were plausible counterexamples to the transitivity of the part relation, that would only show that what is at issue is something other than the mereological part relation. Those pioneering the use of mereology didn't take it upon themselves to provide an analysis accommodating all the nuances which can be conveyed with the word "part". That in no way detracts from the use of mereology in the elucidation of concepts where it is clearly applicable, such as the theories of temporal intervals, regions of space, quantities of matter and processes to be explored presently. As it stands, classical mereology is an incomplete theory, and there is little sense in making it even weaker in the attempt to provide an analysis accommodating all the nuances which might in various contexts be conveyed with the word "part". It might be said that such usage should be clearly flagged by speaking of mereological parthood rather than simply of parthood. But this would be unnecessarily cumbersome here since I shall only be speaking of mereological parthood when speaking of parts. In any case, in view of the difficulty of finding convincing counterexamples to transitivity, it is the non-transitive usage which should really be flagged, for example by finding another word for the concept in question. This is the policy adopted here, where in a later chapter the part-like notion of constitution is taken to be a triadic relation between an individual (such as a biological organism), a quantity of matter and a time. Mereology is involved in elucidating this relation in virtue of the mereological relations and operations sustained by some of the relata, namely quantities and times. But the relation itself calls for a somewhat more complicated analysis than simply taking it to be a mereological relation.[4]

Another matter on which critics take issue with traditional mereology is the principle of so-called unrestricted summation. This principle sanctions, so the critics say, the existence of the mereological sum of anything that exists. We have a man and his wife, and so by the principle the sum of the two. But what is that? A man and his wife are humans, but hardly their sum—this comprises independently moveable parts, has two heads rather than the usual one, and so on. As with the putative counterexamples to transitivity, however, this is a travesty of the principle at issue. Summation is an operation generating an object of the same general kind as that on which the operation acts. The operation of addition generates numbers from numbers, the operation of differentiation generates a function from functions and the operation of disjunction generates a sentence from sentences. So with mereology, a mereological operation generates an entity of the same kind as those on which it operates. Accordingly, if the putative sum of a man and his wife is not a human, then the operation of mereological summation is not applicable to humans. Humans do not sustain mereological relations and operations.

An axiom for the real number system states

$$\forall x \, \exists y \, (x + y = 0).$$

No one would take seriously the objection that there is nothing that could be added to Socrates to generate zero. The axiom is understood to apply to a restricted domain of objects, which mathematicians make explicit, for example by first writing "To each pair of real numbers x, y there is associated a real number, denoted by $x + y$, called the *sum* of x and y," and then for the above axiom, "If x is an element of **R**, then there exists an element $y \in \mathbf{R}$ such that $x + y = y + x = 0$" (Lang 1969, p. 15). When arguing above that the sum of two humans is not a human it was implicitly assumed that the universe of discourse (comprising everything of the general kind at issue) was humans. Were it animals or Englishmen, the corresponding point would be that two animals or two Englishmen cannot be summed to yield an animal or an Englishman, respectively. Any application of the formal theory of mereology similarly has a domain of application, so that universal and existential quantifiers range over this domain. But as already indicated, the domain of objects of interest here includes objects not subject to mereological operations, so there is no general axiom to the effect that to each pair of objects there is an associated object which is their sum, still less that for all objects satisfying a predicate φ, there is a sum of all and only those objects.

There are several ways of capturing the restriction. One would be to introduce a first-order language with the one-place predicates "quantity (of matter)", "(spatial) region", "(interval of) time" and "individual" and lay down axioms to the effect

[4] Those who argue that constitution is identity will complain that this is moving the goalposts. But I reply that they set up their discussion by imposing their construal from the outset, and only consider compliant examples.

that the sum of any things satisfying the first predicate exists, the sum of any things satisfying the second predicate exists and the sum of any things satisfying the third predicate, together with a further condition to be discussed presently, exists. There would be no axiom providing for the existence of sums of individuals. Another way of proceeding might be to lay down the existence of sums of objects which are not individuals. There is nothing in principle wrong with allowing entities which are, say, the sum of a time and a spatial region, or of a spatial region and a quantity of matter, or of all three. But these thingummybobs are not times, nor are they spatial regions nor quantities of matter. What would have to be done is to specifically distinguish those sums of times as times, those sums of regions as regions and those sums of quantities as quantities from thingummybobs in general. But since I have nothing to say about thingummybobs, the primarily expository account developed here will be facilitated by circumventing such things altogether.

The procedure adopted here is to regiment with a many-sorted language, reserving the familiar variables $x, y, z, \ldots$ for individuals, and introducing the variables $t, t', t'', \ldots, t_1, t_2 \ldots$ for times, the variables $p, q, r, \ldots, p_1, p_2 \ldots$ for regions of space and the variables $\pi, \rho, \sigma, \ldots$ for quantities of matter. Finally, processes are referred to by variables $e, f, g, \ldots$. The mereological axioms governing each kind of entity to which they apply are laid down separately. This ensures that sums do not sum over entities from several different categories. The procedure is illustrated by laying down the classical axioms of mereology for quantities of matter, following Leonard and Goodman (1940) except that their use of set theory is avoided by confining the axiomatisation to a first-order theory with an axiom schema.

Using "|" for the mereological predicate "is separate from", which is taken as primitive, the part relation, written $\subseteq$, is defined by

$$\text{Df}\subseteq \quad \pi \subseteq \rho \;\equiv\; \forall\sigma\,(\sigma \mid \rho \supset \sigma \mid \pi).$$

This part relation is to be distinguished from the notion of a proper part, written $\subset$ and defined by

$$\text{Df}\subset \quad \pi \subset \rho \;\equiv .\; \pi \subseteq \rho \wedge \pi \neq \rho.$$

Overlapping, or having a common part, $\circ$, is defined by

$$\text{Df}\circ \quad \pi \circ \rho \;\equiv\; \exists\sigma\,(\sigma \subseteq \pi \wedge \sigma \subseteq \rho).$$

There are infinitely many axioms, namely the two sentences

$$\text{MQ1} \quad \forall\pi\,\forall\rho\,(\pi \subseteq \rho \wedge \rho \subseteq \pi \,.\!\supset\, \pi = \rho),$$

specifying the criterion of identity for quantities, and

$$\text{MQ2} \quad \forall\pi\,\forall\rho\,(\pi \circ \rho \;\equiv\; \sim(\pi \mid \rho)),$$

governing the mereological relations, together with all sentences of the kind

$$\text{MQ3} \quad \exists\pi\,\varphi \supset \exists\rho\,\forall\sigma\,(\sigma \mid \rho \;\equiv\; \forall\pi(\varphi \supset \sigma \mid \pi)),$$

where φ is a formula. This last postulate is not a single formula but an axiom-schema, representing every formula formed from the schema by replacing φ with a definite formula. It provides for the existence of sums. Uniqueness follows, permitting the definition of the sum of φ-ers, $\Sigma\pi\ \varphi(\pi)$, by

$$\text{Df}\ \Sigma\quad \exists\pi\ \varphi(\pi) \supset . \sigma = \Sigma\rho\,\varphi(\rho) \quad\equiv\quad \forall\pi\ (\pi \mid \sigma \ \equiv\ \forall\rho\,(\varphi(\rho) \supset \pi \mid \rho)).$$

Because there is no null element, the sum of φ-ers is not defined unless there is at least one φ-er; hence the existential condition in MQ3 and Df Σ. The binary sum, $\pi \cup \rho$, is defined by the condition "is identical with π or ρ":

$$\text{Df}\cup\quad \pi \cup \rho = \Sigma\sigma\ (\sigma = \pi \vee \sigma = \rho),$$

existence in this case being provided for by the theorem of predicate logic: $\exists\sigma\ (\sigma = \pi \vee \sigma = \rho)$.

These axioms are adequate to establish the classical properties of the mereological relations, such as transitivity and reflexivity for $\subseteq$, transitivity and irreflexivity for $\subset$, symmetry and irreflexivity for $\mid$, symmetry and reflexivity for $\circ$ and classical principles such as

$$\pi \not\subseteq \rho \ \equiv\ \exists\sigma\ (\sigma \subseteq \pi \wedge \sigma \mid \rho),$$

where $\pi \not\subseteq \rho$ abbreviates $\sim(\pi \subseteq \rho)$. Simons (1987, p. 29) calls the left-to-right implication here the strong supplementation principle. It couldn't be satisfied if there were a null element, which would trivialise separation and overlap by being a part of every quantity and entail that nothing would be separate from anything else (everything overlaps everything). Thus, a quantity π not part of the null element would, by the principle, have a part separate from the null element; but there could be no such part.

The unrestricted or general sum, $\Sigma\pi\ \varphi(\pi)$, is of course restricted (in the larger context of a many-sorted system) in the sense that only quantities are incorporated in the summation. It is unrestricted in the sense that it is defined for any predicate of quantities, φ, provided at least something satisfies it. Summation appropriate for times will not be unrestricted in this sense. There is an allied operation deriving from the notion of part, which is an ordering relation (transitive, antisymmetric—i.e. MQ1—and reflexive), namely that of being the least upper bound of the φ-ers, including all and only the parts of the φ-ers. This is symbolised as $\Lambda\pi\ \varphi(\pi)$ and defined by

$$\text{Df}\ \Lambda\quad \exists\pi\ \varphi(\pi) \supset . \pi = \Lambda\rho\,\varphi(\rho) \quad\equiv\quad \forall\rho\,(\pi \subseteq \rho \ \equiv\ \forall\sigma\ (\varphi(\sigma) \supset \sigma \subseteq \rho)).$$

It can be shown that the unrestricted sum $\Sigma\pi\ \varphi(\pi)$ is identical with the corresponding least upper bound, $\Lambda\pi\ \varphi(\pi)$.

An unrestricted product operation, the product of φ-ers, written $\Pi\pi\ \varphi(\pi)$, can be defined in terms of the sum operation, as the sum of the parts common to all the φ-ers. A prerequisite ensuring the existence of a common part must be satisfied:

$$\text{Df}\ \Pi\quad \exists\pi\ \forall\rho\ (\varphi(\rho) \supset \pi \subseteq \rho) \ \supset\ \Pi\rho\,\varphi(\rho) = \Sigma\pi\ \forall\rho\ (\varphi(\rho) \supset \pi \subseteq \rho).$$

Existence and uniqueness of the product are then an immediate consequence of the existence and uniqueness of sums. Again, where the predicate φ is simply "is identical with π or ρ" the product in question is the binary product of π and ρ, written $\pi \cap \rho$.

The difference between two elements π and ρ introduces a third element, which can again be defined in terms of the sum operation as the sum of all those parts of π separate from ρ. The existential prerequisite is therefore $\exists\sigma\,(\sigma \subseteq \pi \wedge \sigma \mid \rho)$. But this is equivalent with $\pi \nsubseteq \rho$ by the equivalence mentioned above involving the strong supplementation principle and its converse. Accordingly, the difference operation, $-$, is defined by:

$$\text{Df}- \quad \pi \nsubseteq \rho \supset \pi - \rho = \Sigma\sigma\,(\sigma \subseteq \pi \wedge \sigma \mid \rho).$$

Finally, although the axiom MQ3 doesn't allow for the existence of the sum of things satisfying an impossible condition, such as being not self identical, which would yield a null element, it does allow the existence of the sum of things satisfied by everything, such as being self-identical. The corresponding sum is called the or the universe of quantities, U_q, formally defined by:

$$\text{Df}\,U_q \quad U_q = \Sigma\pi\,(\pi = \pi).$$

The complement of a quantity π, written $-\pi$, can then be defined as $U_q - \pi$ provided $\pi \neq U_q$. In this case the prerequisite for the existence of the difference is always satisfied for any quantity, π, except U_q.

The purely mereological features introduced in this section are not sufficient to bring out characteristic properties of quantities of matter that distinguish them from other entities subject to mereological relations and operations. For that purpose, further vocabulary will be introduced presently relating quantities to other kinds of entity in characteristic fashion. In the case of times, supplementing the mereological principles with further axioms without further primitives goes a long way to delimiting what kind of entity they are.

1.3 Times

There are certainly extended times or intervals. The times explicitly or implicitly required for the description of the features and behaviour of ordinary, middle-sized objects and quantum objects are intervals. Intervals are the times needed to sustain the properties and relations of continuants and the times it takes processes to occur: colliding, combustion and erosion take time. Various arguments from science can be given that macroscopic bodies take time to change state and cannot do so instantaneously. If a projectile were to change direction instantaneously when rebounding from a wall, for example, infinite forces would be required. Special relativity requires time for the effect of an impact stopping a body to be conveyed from the front to the back. Again, all the atoms constituting a moving body, even a solid one, are in motion relative to the centre of mass, and it is their time-averaged motion that constitutes an overall motion of the macroscopic body. The same could be said of any change of state of a body involving, for example, a change of temperature or of phase. Time averages and relaxation times for the re-establishment of equilibrium presuppose intervals of time. Aristotle contemplated the paradox arising from the idea that time comprises a series of instants, which would seem to require that a body starting to move either has a last instant of immobility and no first instant of motion or no last moment of rest and a first of motion, with no grounds for determining which. Hamblin (1969, 1971) proposed an analysis of time as comprising intervals and not instants, in terms of which the question doesn't arise.

Whether there are any good positive arguments for the existence of instants is by no means clear. It has been suggested that if intervals are construed as sets then the distinction between open and closed intervals can be used to account for Vendler's (1967) distinction between activities and accomplishments (described by the use of the preterit tense in Spanish), thus providing motivation for the introduction of boundary instants. On this view, activities predicate open intervals, whilst accomplishments predicate closed intervals. This way of marking the distinction is highly implausible as a semantic analysis and an alternative account can be given without resorting to instants once an account of intervals is available which doesn't treat them as sets of instants. An activity predication is distributive (it holds of every subinterval of a time at which it holds) and cumulative (if it holds of two overlapping or abutting times, then it holds of their sum). An accomplishment predication of time, on the other hand, holds for exactly the time at issue, and not for any sub- or superinterval.[5]

Hamblin's approach takes a dyadic relation he calls abutment as primitive, construed as asymmetric with axioms ensuring it is irreflexive and transitive, in terms of which a dyadic earlier than relation and mereological concepts such as the overlapping relation and dyadic operations of "intersection" and "join" are defined. Following the spirit but not the letter of Hamblin's proposal, the system outlined here (essentially that of Needham 1981) is based on mereological primitives, in terms of which a *symmetric* dyadic relation of abutment is defined as well as a triadic relation of betweenness. The possibility of doing this depends upon appropriate axioms, which when added to the basic mereological axioms ensure the adequacy of the definitions. It can be shown that a dyadic order relation is not definable in terms of these concepts. This leaves it open to supplant the theory with something providing the basis for the definition of a direction of time and a dyadic earlier than relation, as discussed for example in the following subsection.

Taking mereological separation as the sole primitive, two standard mereological axioms are laid down by analogy with MQ1 and MQ2 of the last section:

$$\text{MT1} \quad \forall t_1 \forall t_2 (t_1 \subseteq t_2 \wedge t_2 \subseteq t_1 \;.\!\supset\; t_1 = t_2)$$

$$\text{MT2} \quad \forall t_1 \forall t_2 (t_1 \circ t_2 \;\equiv \sim (t_1 \mid t_2)),$$

where part and overlapping, as well as proper part, are defined as in the last section. The relation of abutment, A, can then be defined in purely mereological terms, to

[5]Extending the mereological base by accommodating distinctions of open and closed intervals in the manner of Clarke (1981, 1985) and as developed by Randlee et al. (1992) will not, therefore, be of interest here.

the effect that abutting intervals are overlapped by some interval every overlapping interval of which overlaps one or other of two intervals at issue. Formally,

$$\text{Df. } A \quad At_1t_2 \;\equiv .\; t_1 \mid t_2 \wedge \exists t_3\, (t_3 \circ t_1 \wedge t_3 \circ t_2 \wedge \forall t_4\, (t_4 \circ t_3 \;\supset .\; t_4 \circ t_1 \vee t_4 \circ t_2)).$$

It will be convenient to say that two intervals are connected, C, if they either abut or overlap; formally,

$$\text{Df. } C \quad Ct_1t_2 \;\equiv .\; At_1t_2 \vee t_1 \circ t_2.$$

Distant intervals, that neither abut nor overlap, are said to be *disconnected*.

Intervals don't have gaps. The time stretching from noon to 2 o'clock and then from 6 o'clock to 8 o'clock is not an interval; it may be possible to interpret the expression "the time" here as a measure totalling in this case 4 h, but it is not a time, which must be unbroken. On this understanding, there is no sum of the time stretching from noon to 2 o'clock and that stretching from 6 o'clock to 8 o'clock, as there would be if an unrestricted general sum operation were introduced like that governing quantities. A restriction on summation is required ensuring that there are no such interrupted times. Accordingly, summation is introduced in the form of a binary operation, $\cup$, by first laying down the following existence axiom:

$$\text{MT3} \quad Ct_1t_2 \supset \exists t_3\, \forall t_4\, (t_4 \mid t_3 \;\equiv .\; t_4 \mid t_1 \wedge t_4 \mid t_2),$$

after which uniqueness is readily established, allowing the definition of a dyadic sum operation as follows

$$\text{Df. } \cup \quad Ct_1t_2 \;\supset .\; t_3 = t_1 \cup t_2 \;\equiv\; \forall t_4\, (t_4 \mid t_3 \;\equiv .\; t_4 \mid t_1 \wedge t_4 \mid t_2).$$

In the absence of a general sum operation, a universal time cannot be defined by a universal condition as a universal quantity was defined, and the possibility of a finite iteration of the dyadic operator lapping up all the times in a universal time is precluded by later axioms ensuring the infinite extent of time.

With no general sum, the binary product cannot be defined as a sum, and separate provision must be made for the product operation with an appropriate axiom for existence (provided the intervals in question overlap), from which uniqueness is readily established allowing the dyadic product, $\cap$, to be defined in the usual way:

$$\text{MT4} \quad t_1 \circ t_2 \supset \exists t_3\, \forall t_4\, (t_4 \mid t_3 \;\equiv .\; t_4 \mid t_2 \vee t_4 \mid t_2)$$

$$\text{Df. } \cap \quad t_2 \circ t_2 \;\supset .\; t_3 = t_2 \cap t_2 \;\equiv\; \forall t_4\, (t_4 \mid t_3 \;\equiv .\; t_4 \mid t_2 \vee t_4 \mid t_2).$$

Difference, $-$, can be defined along essentially the usual lines, except that precautions must be taken to ensure that intervals with holes are not generated by imposing a restriction confining subtraction to an interior end part. The relation of being an interior end part to a time, denoted by E, is in turn defined in terms of being both a proper part and standing in the relation of internal abutment, I, the latter defined in terms of the existence of a common abutting interval:

Df. I $It_1t_2 \equiv. t_1 \circ t_2 \wedge \exists t_3 (At_3t_1 \wedge At_3t_2)$.

Df. E $Et_1t_2 \equiv. It_1t_2 \wedge t_1 \subset t_2$.

Df. $-$ $Et_1t_2 \supset. t_3 = t_2 - t_1 \equiv \forall t_4 (t_4 \subseteq t_3 \equiv. t_4 \subseteq t_2 \wedge t_4 \mid t_1)$.

A betweenness relation, B, can be defined and standard betweenness axioms (MT5—MT7 below) laid down ensuring a linear order to time. The general idea of the definition is that t_2 lies between t_1 and t_3 if everything including all of t_1 but no part of t_2 is separate from everything including all of t_3 but no part of t_2. This definition wouldn't be adequate if time were circular (or had the form of a figure "8", or other closed forms), because t_2 might lie between t_1 and t_3 and intervals including t_1 and t_3, respectively, might overlap despite each being separate from t_2. In the present case, the possibility of circular time is precluded by the axiom MT6.[6]

Df. B $Bt_1t_2t_3 \equiv. t_1 \mid t_2 \wedge t_2 \mid t_3 \wedge \forall t_5 \forall t_4 (t_1 \subseteq t_5 \wedge t_3 \subseteq t_4 \wedge t_5 \mid t_2 \wedge t_4 \mid t_2 .\supset t_5 \mid t_4)$

MT5 $t_1 \mid t_2 \wedge t_2 \mid t_3 \wedge t_3 \mid t_1 .\supset. Bt_1t_2t_3 \vee Bt_1t_3t_2 \vee Bt_2t_1t_3$

MT6 $\sim (Bt_1t_2t_3 \wedge Bt_1t_3t_2)$

MT7 $Bt_1t_2t_3 \wedge t_4 \mid t_1 \wedge t_4 \mid t_2 \wedge t_4 \mid t_3 .\supset. Bt_1t_2t_4 \vee Bt_4t_2t_3$.

The properties $Bt_1t_2t_3 \supset Bt_3t_2t_1$ (1–3 symmetry) and $Bt_1t_2t_3 \supset t_1 \mid t_3$ are immediate consequences of the definition of betweenness. It also follows from the definition of betweenness and the existence of binary sums that abutment implies the absence of any intervening interval.

Further axioms can be laid down. What might be called the unity of time, that intervals always have abutting neighbours, is ensured by

MT8 $t_1 \mid t_3 \supset. At_1t_3 \vee \exists t_2 (Bt_1t_2t_3 \wedge At_1t_2 \wedge At_2t_3)$.

Time's infinite extent in both directions is expressed by

MT9 $\exists t_1 \exists t_3 Bt_1t_2t_3$.

In another context, this might be considered compatible with time being circular, and so not implying time's infinite extent. Euclid's second postulate, that a straight line can be produced continuously in a straight line, is ambiguous between merely saying that a line is boundless—there is always another point further along, without end—and asserting that a line is infinite in extent. But interpreting it in the manner of MT9 allows it to be satisfied by the points along a great circle on the surface of a sphere, or the straight lines of Riemann geometry. In the present context, the other axioms require times to be intervals with parts and not structureless instants.

[6]An appropriate order relation for circular time (or time in an incomplete theory which leaves the question of linearity or circularity open) would be a 4-place relation of separation closure, which could be defined on the basis of the same mereological resources but will not be pursued here (see Needham 1981, p. 57).

Intervals along a circle can be successively summed with their neighbours so as to finally incorporate the entire circle in a single time which can't be flanked by others, in contradiction with MT9. In any case, if time were circular $Bt_1t_2t_3$ would imply $Bt_1t_3t_2$, which is precluded by MT6. The latter axiom is not therefore independent of the other axioms.

Finally, with the addition of an axiom expressing the infinite divisibility of time,

MT10 $\exists t_1\ t_1 \subset t_2$,

the axiomatisation of intervals is complete. That is to say, considered as a first-order theory dealing only with times, the theory described in this section is complete Needham (1981, pp. 60–2). Other definite alternatives (such as the existence of a single endpoint while the other end remains open instead of MT9) yield alternative complete systems, but dropping axioms on which we might not want to make a stand leaves the axiomatisation incomplete.

We have now seen how a restricted sum operation can be introduced with the consequence that there is no universal time. A further consequence is that the least upper bound of two intervals is not necessarily their sum. There is no sum of the two times stretching from noon to 2 o'clock and from 6 o'clock to 8 o'clock, whose least upper bound is the interval stretching from noon to 8 o'clock. The restriction on the summation of intervals has repercussions on the summation of processes, which are discussed in Chap. 8.

Supplementing mereology in this way has given us a theory whose entities can be construed as times standing in a linear order. The only primitive is a mereological one. It is therefore hardly surprising that the ordering is undirected, like that along a spatial dimension. A simple application of Padoa's method shows that an asymmetric dyadic order relation cannot be defined with the resources of this theory. Conventionally assigning one of two selected times as the earlier of the two suffices to impose a direction on the betweenness ordering that can be expressed as a general definition of the earlier than relation. But there is a view which sees the direction of time as having its origin in the naturally occurring causal processes in the world. The mereological theme of this chapter is interrupted with a bare sketch of how this idea might be implemented in the present context in the following subsection.

1.3.1 Directed Time

Suppose t_1, t_2, t_3, t_4 are pairwise distinct (non-overlapping) times, and consider the two pairs (t_1, t_2), (t_3, t_4). The pairs could be said to have the same direction if all four times were temporally betweenness ordered as the corresponding variables are spatially betweenness ordered when displayed, for example, thus: t_1, t_2, t_3, t_4, but not when, for example, displayed thus: t_2, t_1, t_3, t_4. There are 12 (4! / 2) ways of arranging four pairwise distinct intervals t_1, t_2, t_3, t_4 on a line, six of which correspond to the pair (t_1, t_2) having the same direction as the pair (t_3, t_4). A general definition of having the same direction in terms of these six alternatives can be conveniently formulated by first defining the four-place relation $B(t_1, t_2, t_3, t_4)$ by

$$B(t_1, t_2, t_3, t_4) \equiv . \; B(t_1, t_2, t_3) \wedge B(t_2, t_3, t_4).$$

The four-place relation $SD(t_1, t_2, t_3, t_4)$, for "the pair t_1, t_2 has the same direction as the pair t_3, t_4" is then defined by

$$SD(t_1, t_2, t_3, t_4) \equiv . \; B(t_1, t_2, t_3, t_4) \vee B(t_1, t_3, t_2, t_4) \\ \vee B(t_1, t_3, t_4, t_2) \vee B(t_3, t_1, t_2, t_4) \vee B(t_3, t_4, t_1, t_2).$$

The general idea of the causal theory of the direction of time is that the times in each pair are causally related in some way. On the relational theory of causation, according to which causation takes the form of one event standing in the causal relation as the cause of another, the times in a pair could be the times of occurrence of these events. A pair of times could then be understood as standing in the relation K defined by

$$K(t_1, t_2) \equiv \exists e_1 \exists e_1 \, (Occurs(e_1, t_1) \wedge Occurs(e_2, t_2) \wedge e_1 \, causes \, e_2).$$

A somewhat different view has it that a causal process transforms one state into another, and the times of our pairs would be the times when the states obtain. With a function $T(s)$ assigning a time when a state s obtains, the relation K might then be defined along the lines

$$K(t_1, t_2) \equiv \exists e \exists s_1 \exists s_1 \, (Tr(e, s_1, s_2) \wedge t_1 \subseteq T(s_1) \wedge t_2 \subseteq T(s_2)),$$

where $Tr(e, s_1, s_2)$ is understood to say that process e transforms state s_1 into state s_2. At all events, with some such relation between a pair of times, the contention of the causal theory is that an empirical generalisation holds along the lines of

> For any four pairwise distinct (separate) times t_1, t_2, t_3, t_4 it holds that: if $K(t_1, t_2) \wedge K(t_3, t_4)$ then $SD(t_1, t_2, t_3, t_4)$,

or perhaps a slightly weaker statistical generalisation

> For nearly all four pairwise distinct times t_1, t_2, t_3, t_4 it holds that: if $K(t_1, t_2) \wedge K(t_3, t_4)$ then $SD(t_1, t_2, t_3, t_4)$,

which allows for the occasional case of backwards causation.

The earlier than relation serves as the basis for determining the direction of causation in the regularity theory of causation, where the relation of constant conjunction does not itself suffice to provide an asymmetric relation. Even those who don't adhere to the regularity theory may wonder how it is possible to distinguish cause from effect, and the problem may be of independent interest for those who allow cause and effect to be simultaneous.

Reichenbach addressed this problem of finding an independent criterion of causal priority in his early causal theory of time, which was more ambitious than the present theory and sought to provide a causal basis of the temporal ordering itself and not merely the direction of time. He had the idea of distinguishing the cause by making a small modification or mark. If it really is the cause that is modified, then the mark will be transferred to the effect, whereas if the event marked is the effect, then there will be no corresponding modification of the cause. In other words, if events of the kind X cause events of the kind Y, and "*" indicates a variation, then only event pairs of the following kinds are observed:

$$\{X, Y\}, \ \{X^*, Y^*\}, \ \{X, Y^*\},$$

and never event pairs of the kind:

$$\{X^*, Y\}.$$

However, it was objected that we might well observe the following six events:

$$x_1, \ y_1, \ x_2^*, \ y_2^*, \ x_3, \ y_3^*,$$

where x_1 is of type X, x_2^* is of type X^*, and so forth. Reichenbach wants us to interpret this sequence in such a way that it comprises the three pairs

$$\{x_1, y_1\}, \ \{x_2^*, y_2^*\}, \ \{x_3, y_3^*\},$$

and the pair $\{x_2^*, y_1\}$ doesn't occur. But why should we do this if we don't already take for granted that there is something corresponding to the left-to-right ordering in the sequence? If we know no more than that the six events occur, we could pair them in the following way:

$$\{x_2, y_1\}, \ \{x_3, y_2^*\}, \ \{x_1, y_3^*\}.$$

In other words, the mark method is circular!

With the lesser ambition of seeking a causal basis merely for the direction of time rather than the entire order opens up a way of circumventing the objection and using the mark method as a criterion of causal priority (Needham 1985, pp. 167–9; Faye 1997 , pp. 221–2). Suppose that the following six events are temporally betweenness ordered in such a way as to correspond to the spatial betweenness ordering of the letters:

$$x_1, \ y_1, \ x_2^*, \ y_2^*, \ x_3, \ y_3^*.$$

Choosing of the desired pairs without presupposing the direction of time is possible by considering the interval between the events in a sufficiently large number of similar cases. Making the assumption that all causally connected events x, y of type $\{X, Y\}$ are always temporally separated by equally long intervals of time, a sufficiently large number of observations would exhibit constant (the same) interval lengths only between pairs of events of the following kinds:

$$\{X, Y\}, \ \{X^*, Y^*\}, \ \{X, Y^*\},$$

and the unwanted pairs of the kind

$$\{X^*, Y\}$$

would be excluded. Presupposing a betweenness ordering weakens the original ambition of the causal theorists, but provides for an approach to causal priority adequate for present purposes.

An alternative approach to the direction of time is based on the idea of increasing entropy along the lines formulated by Denbigh and Denbigh in the following passage

> Consider a number of macroscopic systems [bodies, possibly a system of several bodies] A, B, C, etc., each of them adiabatically isolated. Let t_i, t_j, t_k, etc., refer to a sequence of clock readings, but without reference to which direction of the sequence is the direction of 'later than' or 'the future'. Further, let $S^A_{t_i}$, etc., be the measured entropy (relative to some standard state) of system A, etc., at the instant t_i.
>
> Now the empirical content of the Second Law can be expressed as follows:
>
> $$\text{If } S^A_{t_i} \geq S^A_{t_j} \text{ then } S^B_{t_i} \geq S^B_{t_j}$$
>
> or
>
> $$\text{if } S^A_{t_i} \leq S^A_{t_j} \text{ then } S^B_{t_i} \leq S^B_{t_j}$$
>
> Therefore, in either case, we have:
>
> $$(S^A_{t_i} - S^A_{t_j})(S^B_{t_i} - S^B_{t_j}) \geq 0$$
>
> since the brackets have *the same* sign whether this be positive or negative. A similar reasoning can then be applied to any other system, B, C, D, etc., considered pairwise, and also to any pair of clock times. (Denbigh and Denbigh 1985, p. 16)

The second proposal above for the definition of the relation $K(t_1, t_2)$ is well-suited for adapting this suggestion.

1.4 Regions of Space

A theory of spaces based on mereological principles is developed in this section, analogous to the theory of times just presented although it will be necessary to introduce a specifically spatial primitive predicate in addition to the mereological primitive of separation. On the other hand, it will not be necessary to introduce the same kind of restriction on the underlying mereological sum operation that was essential to the exclusion of unconnected intervals and the adequacy of the definitions of abutment and betweenness as they apply to intervals. Unconnected regions are recognised with a view to allowing them to be occupied by scattered quantities of matter. This leaves it open whether there is a universal region including all of space, and a corresponding universal sum operation. It does seem natural to follow the same course and say that just as time itself is not an interval, so space itself is not a region. As with time, a general sum operation in conjunction with an axiom ensuring the indefinite extension of space (MS8 below) would entail the existence of an infinite region, whereas there was no such corollary to the existence of general sums of quantities. Spatial regions provide possible locations of quantities of matter, which are macroscopic objects occupying finite regions, and although the possibility of there being regions larger than that occupied by any quantity of matter is not to be discounted, no quantity could occupy the whole of infinitely extended space. Accordingly, dyadic operations of sum and product will suffice for present purposes, and there is no universal sum operation.

With variables, $p, q, r, \ldots$ ranging over regions of space, the principles of classical mereology governing mereological relations—analogues of MT1 and MT2—are adopted:

MS1 $\forall p \forall q (p \subseteq q \wedge q \subseteq p \;.\!\supset\; p = q)$

MS2 $\forall p \forall q (p \circ q \;\equiv\; \sim (p \mid q))$.

Although there will be call for a notion of connected regions, there is no general restriction on sums of regions analogous to that on sums of times precluding disconnected regions. General sums of regions might, therefore, be allowed unless it is thought inappropriate that the whole of infinite space should be counted a region. Without any strong reason for disallowing a universal region, but for the sake of formulating a definite system of axioms, summation will be confined to the binary sum operation, $\cup$, introduced by the existence axiom

MS3 $\exists p_3 \forall p_4 (p_4 \mid p_3 \;\equiv.\; p_4 \mid p_1 \wedge p_4 \mid p_2)$.

Uniqueness is easily established and the dyadic sum operation defined by

Df. $\cup$ $p_3 = p_1 \cup p_2 \;\equiv.\; \forall p_4 (p_4 \mid p_3 \;\equiv.\; p_4 \mid p_1 \wedge p_4 \mid p_2)$.

As with times, the product operation must then be introduced by a distinct axiom with a restriction to preclude the existence of null elements,

MS4 $p_1 \circ p_2 \;\supset\; \exists p_3 \forall p_4 (p_4 \mid p_3 \;\equiv.\; p_4 \mid p_1 \vee p_4 \mid p_2)$,

and establishment of uniqueness allows the operation $\cap$ to be defined:

Df. $\cap$ $p_1 \circ p_2 \;\supset.\; p_3 = p_1 \cap p_2 \;\equiv\; \forall p_4 (p_4 \mid p_3 \;\equiv.$
$p_4 \mid p_1 \vee p_4 \mid p_2)$.

There is a specifically spatial primitive, the 1-place predicate Sp, read "p is a stratum". A stratum can be thought of as a region extending infinitely in two directions like a plane of Euclidean geometry, but unlike a plane in so far as it is infinitely divisible into parallel strata (as laid down in axiom MS9 below), i.e. a stratum has "thickness" and is 3-dimensional like any other region. Strata stand in a betweenness relation, B, analogous to the way intervals of time do, defined by

Df. B $Bpqr \;\equiv.\; S[p, q, r] \wedge p \mid q \wedge q \mid r \wedge \forall p_1 \forall r_1 (S[p_1, r_1]$
$\wedge p \subseteq p_1 \wedge r \subseteq r_1 \wedge q \mid p_1, r_1 \;.\!\supset\; p_1 \mid r_1)$.

In order to shorten formulas, the conjunction $Sp \wedge Sq \wedge Sr$ has been abbreviated $S[p,q,r]$. $S[p_1,r_1]$ is understood similarly, and $q \mid p_1, r_1$ abbreviates the conjunction $q \mid p_1 \wedge q \mid r_1$. When these kinds of abbreviation will be used in the sequel, they are always flagged by the occurrence of a comma or square brackets.

Immediate consequences of the definition are that $Bpqr$ implies $Brqp$ and $p \mid r$. The remaining basic principles of order are laid down as axioms:

MS5 $\quad S[p,q,r] \;\wedge\; p \mid q \;\wedge\; q \mid r \;\wedge\; p \mid r \;.\!\supset\!.\; Bpqr \;\vee\; Bqpr \;\vee\; Bprq$

MS6 $\quad \sim(Bpqr \;\wedge\; Bprq)$

MS7 $\quad Bpqr \;\wedge\; Ss \;\wedge\; s \mid p,q,r \;.\!\supset\!.\; Bpqs \;\vee\; Bsqr.$

As with times, further questions arise about the structure of spaces that are not easily decided. Here it is simply assumed (with a view to making the axiomatisation complete) that there is no end to a series of betweenness-ordered strata:

MS8 $\quad S[p,q] \;\wedge\; p \mid q \;.\!\supset\; \exists r\, Bpqr.$

In view of the 1–3 symmetry of betweenness, it follows that $Bpqr \supset \exists sBspq$. It is also assumed that strata are infinitely divisible in the sense that each is the sum of two separate strata:

MS9 $\quad Sp \;\supset\; \exists q \exists r\, (S[q,r] \;\wedge\; q \mid r \;\wedge\; p = q \cup r).$

Elements in a series of betweenness-ordered strata are said to be parallel. More generally, being parallel, $\|$, is defined on strata by

Df. $\|$ $\quad p \parallel q \;\equiv.\; S[p,q] \;\wedge\; \exists p_1 \exists q_1 \exists r\, (p_1 \subseteq p \;\wedge\; q_1 \subseteq q \;\wedge\; Bp_1q_1r).$

Symmetry of the parallel relation follows immediately from the definition, and reflexivity for strata follows given MS9 and MS8. A proof of transitivity can be sketched as follows. Assume $p \parallel q \;\wedge\; q \parallel r$. If $p \mid r$ then a stratum in r lies between one in p and some other not overlapping either introduced directly by the assumption, or by applications of MS8. If p and r are not separate they overlap, and MS9 will provide the parts of each which stand in the betweenness relation to something by MS8. An axiom expressing the unity of space can now be written

MS10 $\quad p \parallel q \;\supset\; \exists r\, (Sr \;\wedge\; p,q \subseteq r).$

A stratum which is not parallel to a given stratum will be orthogonal to it, where orthogonality, #, is defined by

Df. # $\quad p\#q \;\equiv.\; S[p,q] \;\wedge\; \sim(p \parallel q).$

Symmetry of orthogonality follows from that of the parallel relation. Further, orthogonality implies overlapping by MS8 and classical mereology (MS2). Now, there are in fact at least three orthogonal series of strata, the strata in each series being parallel to one of three mutually orthogonal strata which exist in accordance with the axiom

MS11 $\quad \exists p \exists q \exists r\, (p\#q \;\wedge\; q\#r \;\wedge\; p\#r).$

Moreover, it is a feature of space that three mutually orthogonal strata will not only overlap pairwise, as already implied; they will overlap in a region common to all three:

MS12 $p\#q \wedge q\#r \wedge p\#r \;.\!\supset\; \exists s\,(s \subseteq p, q, r)$.

The largest such region common to three mutually orthogonal strata (i.e. the product or intersection of the three) is called a box, described by the predicate *Box*:

Df. *Box* $Box(s) \equiv \exists p\, \exists q\, \exists r\, (p\#q \wedge q\#r \wedge p\#r \wedge \forall u\, (u \subseteq s \equiv u \subseteq p, q, r))$.

With this concept of a box available it is now possible to define a notion of abutment, *A* along broadly the same lines that abutment was defined for temporal intervals:

Df. *A* $Apq \;\equiv.\; p \mid q \wedge \exists r\, (Box(r) \wedge r \circ p \wedge r \circ q \wedge \forall s\, (s \circ r \;\supset.\; s \circ p \vee s \circ q))$.

This allows that either of two regions *connected with one another* may be scattered, understanding "connected" as a 2-place predicate, *C*, analogous to the relation of being connected defined, like the analogous relation on times in the last section, as abutting or ovelapping. In particular, every region is connected with itself. There is another notion of connectedness, expressed by a 1-place predicate, of *being a connected region* with the sense of not being scattered. A connected region, described by the predicate *Conn*, is defined as one that can't be exhaustively partitioned into non-abutting regions:

Df. *Conn* $Conn(p) \equiv \forall q\, \forall r\, (q \mid r \wedge p = q \cup r \;.\!\supset Aqr)$.

Thus although regions, unlike intervals, are not in general connected, some of them are.

Like the theory of times, the first-order theory of regions presented in this section can be shown to be complete by the same general method. The proof is outlined in the following subsection.

1.4.1 Completeness of the Theory of Regions

Abstracting from the regimentation project based on a many-sorted language introduced earlier, the theory of spatial regions presented in the foregoing section can be considered a first-order theory based on the two primitives "|" and "*S*". In view of axioms such as MS8 and MS9, there are no finite models of this theory. It can therefore be shown to be complete by establishing $\aleph_0$-categoricity (Vaught's theorem), for which purpose a denumerable model is described based on the rational numbers. The domain of this model comprises the following sets of triples of rational numbers:

(i) For any pair of rational numbers (positive or negative) a_1 and a_2, the sets $\{\langle x, y, z\rangle : a_1 < x < a_2\}$, interpreted as a family, $\mathbf{S}_x$, of parallel strata, each orthogonal to any member of the family, $\mathbf{S}_y$, of parallel strata $\{\langle x, y, z\rangle : b_1 < y < b_2\}$, for any pair of positive or negative rational numbers b_1 and b_2, which are in turn orthogonal to any member of the family, $\mathbf{S}_z$, of parallel strata $\{\langle x, y, z\rangle : c_1 < z < c_2\}$, for any pair of positive or negative rational numbers c_1 and c_2.
(ii) The family $\mathbf{B}$ of boxes $\{\langle x, y, z\rangle : a_1 < x < a_2, b_1 < y < b_2, c_1 < z < c_2\}$.
(iii) Any finite union or non-null intersection of elements drawn from $\mathbf{B}$, $\mathbf{S}_x$, $\mathbf{S}_y$, or $\mathbf{S}_z$.

The interpretation of the predicate S is given in (i). Separation is interpreted as null intersection, in which case overlapping amounts to having some member of the domain in common. Two elements abut if a box overlaps both and is separate from anything separate from either of the two elements in question. Inspection shows that the axioms are all satisfied by the model.

Clearly, the domain is denumerable. Let $m_0, m_1, m_2, \ldots$ be an enumeration of the elements of the domain of this model, and consider an arbitrary denumerable model whose elements can be enumerated thus: $q_0, q_1, q_2, \ldots$. A standard back-and-forth technique of mapping the one structure onto the other establishes that they are isomorphic. The inductive definition of a 1-1 function f from the domain of the standard model to that of the arbitrary model preserving atomic formulas can be outlined as follows.

Put r_0 identical with m_0, and choose $f(r_0)$ to be the first stratum in the enumeration $q_0, q_1, q_2, \ldots$ if r_0 is a stratum; otherwise, let $f(r_0)$ be the first member of $q_0, q_1, q_2, \ldots$ which is not a stratum. Assume $f(r_0), f(r_1), \ldots, f(r_{n-1})$ have already been defined in such a way that for all i, j from 0 to n–1,

$$r_i \mid r_j \text{ iff } f(r_i) \mid f(r_j), \text{ and}$$

$$Sr_i \text{ iff } Sf(r_i).$$

Then consider the nth assignment. Suppose first that n is odd, and let r_n be the first element in the enumeration $m_0, m_1, m_2, \ldots$ not yet assigned. Consider how r_n is related to $r_0, r_1, \ldots, r_{n-1}$ and assign $f(r_n)$ from the remaining elements of $q_0, q_1, q_2, \ldots$ according to which of the following alternatives holds, ensuring at the same time that if Sr_n, then $f(r_n)$ is chosen so that $Sf(r_n)$, and otherwise, so that not $Sf(r_n)$.

(i) r_n is separate from each of $r_0, r_1, \ldots, r_{n-1}$. Then for $f(r_n)$ choose any element p in the arbitrary model separate from each of $f(r_0), f(r_1), \ldots, f(r_{n-1})$.
(ii) r_n overlaps r_k, for one or more k such that $0 \leq k \leq n-1$. The others are treated as in (i), but for each such k, one of the following alternatives holds:

(a) $r_n = r_k$, so put $f(r_n) = f(r_k)$.
(b) $r_n \subset r_k$, so choose $f(r_n)$ so that $f(r_n) \subset f(r_k)$.
(c) $r_k \subset r_n$, so choose $f(r_n)$ so that $f(r_k) \subset f(r_n)$.

(d) there is a proper part, r, of r_n which is separate from r_k and a proper part, r', of r_k separate from r_n, so choose $f(r_n)$ such that $f(r_n)$ overlaps $f(r_k)$ and there is a proper part, p, of $f(r_n)$ which is separate from $f(r_k)$ and a proper part, p', of $f(r_k)$ which is separate from $f(r_n)$.

If n is even, then consider the first element in the enumeration $q_0, q_1, q_2, \ldots$ not yet assigned and let this element be $f(r_n)$ for some r_n taken from the enumeration $m_0, m_1, m_2, \ldots$ chosen so that Sr_n iff $Sf(r_n)$ and further according to which of the alternatives analogous to (i) and (ii)(a)–(d) holds in respect of how this new element from the arbitrary model is related to $f(r_0), f(r_1), \ldots, f(r_{n-1})$.

The axioms ensure that these choices are always available. Clearly, the induction shows that f is a 1-1 function and for elements, m_i, m_j, of the standard model,

$$m_i \mid m_j \text{ iff } f(m_i) \mid f(m_j), \text{ and}$$

$$Sm_i \text{ iff } Sf(m_i),$$

so that the two models are isomorphic. Hence, the theory of spatial regions is complete.

1.5 Taking Stock

The paucity of the bare mereological principles is apparent from these latter developments of theories of regions and times. Quantities were allowed a general notion of summation whereas summation of times and regions was restricted to a binary sum operation. But nothing like the development from the mereological basis that serves to characterise times as extended, linearly ordered entities in a complete theory is forthcoming for quantities. There are no further axioms, over and above the purely mereological ones but confined to the mereological terminology, that would give some positive specification of how quantities differ from times. There won't even be a simple axiom stating the discreteness of matter, nor even a denial of atomism affirming its infinite divisibility. Regions were distinguished from times with the help of an additional, non-mereological primitive in terms of which a complete axiomatisation was formulated. We will have to look towards the introduction of further vocabulary for a positive specification of the features of quantities that distinguish them from those of times and regions. These will not be non-mereological primitives applying solely to quantities, as the non-mereological primitive "stratum" applied to regions, and allowing for the making of a pure theory, much less a complete theory, of quantities analogous to the pure theory of times and the pure theory of regions. They will be relations standing between one or more quantities and one or more of the other kinds of entity referred to in the many-sorted framework introduced here, notably the occupies relation placing quantities and individuals in a region of space at a time, the constitutes relation specifying a quantity's relation to an individual at a time, substance and phase relations relating quantities to times and relations specifying how quantities are involved in processes. The details are developed in the ensuing chapters. But the general idea can be illustrated with the principle

$$\forall \pi \, \forall t \, \exists p \, Occ(\pi, p, t),$$

where *Occ* is the occupying relation, stating that quantities are always somewhere and confined to regions in the present sense.[7]

This approach is at variance with that taken by philosophers who read into the mereological relations a content that might be suggested by English expressions for these relations, but goes beyond what the mereological axioms justify. The policy here is to understand mereology as the theory of the relations of part, separation, etc. and operations of sum, product and difference as determined by the classical axioms specified above. Application is restricted to times, spaces, quantities and processes which don't infringe the classical axioms. Any call on issues going beyond pure mereology raises questions of what additional vocabulary or principle is required, which should be formulated explicitly. This will not issue in a complete theory, and therefore leave several questions unanswered—in some cases raising doubts about the appropriateness of the question. But I hope it will go some way towards clarifying the interrelations between, and therefore the character of, the middle-sized objects of interest here.

Several of the features ascribed to times and regions in the theories presented earlier are ones we can't be certain about. Whether intervals of time and spatial regions are infinitely divisible or ultimately discrete, for example, are controversial issues which should, perhaps, be left open. Much the same might be said about whether they are indefinitely extendable in the sense of MT9 and MS8. This doesn't really detract from the point made at the beginning of this section about the paucity of the purely mereological principles governing quantities, which can be contrasted with a range of alternative complete mereologically-based systems for

[7]What Lowe (1998, p. 96) calls existing in time distinguishing material from abstract objects calls for specific predicates like "occupies". It is not sufficient to have some property at a time, or as Lowe would have it, to have had, to have now or to be going to have a property. Although I wouldn't say that $2 + 2 = 4$ is ever, and certainly not always, true (nor as Arthur Prior would have it, that it always was, is and always will be that $2 + 2 = 4$), numbers do nevertheless stand in relations to times. For example, 4 could well be the present volume of a quantity of gas in cubic meters, or the length in seconds of an interval of time, or the factor by which one interval is longer than another, or the factor by which the length of an interval in seconds is greater than the weight of a quantity of matter in grammes. A time-dependent existence predicate $E(\pi, t)$ expressing existence in time could be defined in terms of a specific predicate like "occupies" as $\exists p \, Occ(\pi, p, t)$. But I wouldn't call the relation between 4 and t defined by "There is a quantity π whose volume in m^3 at t is 4" time-dependent existence expressing the existence of the number 4 in time. The difference regarding existence in time is not merely a matter of being related to times, but specifically that quantities of matter occupy a region of space at a time whereas numbers don't. Other features that presuppose occupying a region of space, such as participating in a process of reacting chemically with another substance, are also marks of existence in time.

times and regions drawn up from alternative sets of axioms, imposing boundedness, for example, instead of unboundedness. It might not even be necessary to go so far as to formulate complete theories of space and time in order to make the point about the little the purely mereological principles say about the nature of quantities.

Meagre though they may be, this is not to say that the mereological principles are without content. They impose constraints on the objects falling within their domain, and any application of mereology must consider whether the objects of interest do in fact comply. Regions and intervals are paradigm cases. Material objects call for a pause for thought. Loosely understood, it seems that sometimes they do, sometimes they don't. Rather than trying to accommodate the recalcitrant cases by bending mereology, a distinction is drawn here between quantities, which do, and individuals, which don't, comply with basic mereological principles. Both may be considered material objects, but it is those which are quantities that are of primary interest here. Chapter 3 will say enough about individuals to make the contrast clear and establish that quantities are not to be confused with individuals. Questions about how individuals are distinguished are pursued up to a point, but without any claim to tackle the finer details of the individuation of, for example, biological organisms. Drawing such a distinction may be at odds with the intuitions of some of those who argue for modifying mereological principles. But I argue that it conforms to intuitions shaped by the elementary science learnt in school which arose in the course of the history of science and was subsequently developed into the more sophisticated modern science of macroscopic objects to deal with the complex behaviour of matter.

Throughout the development of additional principles articulating the notion of a quantity by relating quantities to other kinds of object, the mereological relations and operations retain their strictly algebraic character. Summation, in particular, is an operation that generates entities from entities of the same kind. Its application to quantities is illuminatingly contrasted with the lack of a corresponding operation on individuals (Sect. 3.5). Summation is also to be distinguished from a process of generating a quantity from others and taking place in time, or as arising as a result of parts sticking together. Quantities will be related to processes taking place in time, but not in this way, and questions pursued concerning the relation of mereological relations and operations to the combination of elements in compounds, the interaction of solutes with solvents in solutions and the intermolecular forces underlying the various phases of matter. But there is no Special Composition Question to be answered seeking the kind of cohesion serving as the criterion for "when summation happens" or whatever the appropriate expression might be.

When van Inwagen (1987) raises his Special Composition Question, he considers contact, fastening, cohesion and fusion to defy precise definition and concludes that answers are too vague to justify the existence of sums in a wide variety of cases. But these notions, vague or not, are completely irrelevant to the question of what stands in mereological relations or what the application of a mereological operation refers to. A mixture of oxygen and hydrogen in the ratio of one mole to two at normal pressure and a temperature over 100 °C, say 150 °C, is not at thermodynamic equilibrium, but is to all intents and purposes stable if not subject to short sharp imputs of energy that would start a chain reaction. The mixture is a mereological sum of the quantity of oxygen and the quantity of hydrogen. It is not a compound of these two quantities. That would be induced by, say, a short exposure to a flame, which would result in the combination of hydrogen and oxygen yielding the compound known to us as water. Under the same conditions of temperature and pressure, this mereological sum of hydrogen and oxygen would then occupy two-thirds of the volume occupied by the uncompounded mixture, and conforms to thermodynamic criteria of comprising a single substance[8] (discussed in Sect. 6.4 below). The flame has not induced the formation of a mereological sum from something that wasn't. It has changed the state of one and the same mereological sum from that of being a mixture (a solution) to that of being a compound, so that it now bears the property of being water that it didn't before.

Mereology has nothing to do with the notion of composition, compound formation or any other notion of bodies adhering to one another. As we saw when defining the notion of abutment for times and for spatial regions, even the notion of contact is beyond the scope of mereology and definable only with the addition of further axioms in the one case, and further axioms together with an additional primitive in the other. When we come to deal with the general notion of abutment or contact between quantities and between individuals in the next chapter, we will see that it presupposes all this and more. Being a homogeneous mixture (i.e. a solution), being a compound (daltonide or berthollide), being a colloid, and so on, are chemical concepts involving some kind of interaction which might yield to analysis on a basis that includes purely mereological concepts, but not on these alone. Cohesion is not a mereological concept. Certainly, van Inwagen's Special Composition Question has no bearing on mereology if it literally asks for a physical process, let alone one akin to a human action as suggested by the passage:

> Suppose one had certain non-overlapping material objects, the *x*s ... what would one have to do—what could one do—to get the *x*s to compose something? (van Inwagen 1987, p. 23)

Nor does it if it asks for some state of cohesion. Confusion on this point is best avoided by not speaking of composition when summation[9] is at issue. Accordingly, "summation" is the term used here for the mereological operation; "composition" is reserved for something else.

This might be the place to emphasise that there is no assumption underlying the discussion in the following chapters of the existence of instants in the sense of indivisible atoms of time with no duration. In particular, there is no presumption that a predicate holding of an interval of time (as well, perhaps, of regions and quantities) holds, or depends upon other predicates holding, of any, let alone every,

[8]"Substance", here and throughout this book, is used in its ordinary English sense of "chemical substance".

[9]Or van Inwagen's richer notion of a mereological partition (mutually separate parts whose sum is identical with the whole). In Sect. 9.2 it is argued that a quantity of water cannot be mereologically partitioned into hydrogen and oxygen, nor substances in general into atoms.

instant purportedly making up the interval. The word "interval" is naturally used to describe a time and is so used here, without any commitment to the usage in set theory in the sense of a set of all the elements, usually numbers, lying between two limits along a linearly ordered set-theoretical structure. Unless the contrary is explicitly indicated, any set-theoretical usage of admittedly ambiguous terminology should be assumed not to be at issue. Similarly, regions are not assumed to comprise points and quantities are not assumed to comprise point particles.

The minute composition underlying the continuous matter we are familiar with at the macroscopic level has fascinated philosophers since the times of the ancients. In his interesting recent assessment of discussions of the paradoxes of infinite divisibility of matter in renaissance and enlightenment philosophy, Holden (2004) rejects the standard modern interpretations which treat the issues as mathematical exercises and criticise older writers for their ignorance of the properties of denseness and limits exhibited by rational and real numbers:

> Given this standard interpretation, the paradoxes turn out to be quite easy to disarm. ... [T]hey rest on the period's halting and flawed mathematical understanding of the logic of limits and the infinite series. ... But this mathematical interpretation is quite mistaken ... [T]his sort of mathematical model that considers the form or structure of infinitely divisible body *in abstracto*, divorced from all thought of the actual physical filling or stuffing of real concrete bodies, may indeed be perfectly unproblematic. ... [N]atural philosophers of the period saw the infinite divisibility of *material body* as the key problem, and not the infinite divisibility of space or extension in general. ... Those philosophers, such as the Newtonians, who ... draw a distinction between body on the one hand and space on the other phrase the problem in terms of the divisibility of body, not space. ... But this point is lost in much of the recent commentary. (Holden 2004, pp. 20–1)

The underlying thesis remains as true as ever. The continuity of macroscopic matter is not understood in quasi-mathematical fashion by postulating points in abstraction from considerations of "the actual physical filling or stuffing of real concrete bodies". But the situation is not redressed for the modern reader by "missing premises" of the kind:

> (a) the argument [Zeno's metrical paradox applied to bodies] assumes actual parts, and hence an actual infinity of m[etaphysically]-indivisible ultimate parts [spatially distinct parts which could exist separately—at a distance—from one another]; (b) those ultimate parts are all the same size; and (c) they could not be extensionless—[in view of which] the argument is much more compelling than the recent literature has allowed. (Holden 2004, p. 47)

Nor is there any plausibility in postulating either material points or material extended, partless atoms. Purely mechanical material points are paradoxical: they don't stand in contact and collisions call on extra-material forces. This is not advanced, as Holden (pp. 263–72) notes, by Boscovich's strategy of associating a material point with a force field. The central points are not independent thing-like entities, just centres of the fields of force, and identifying matter exclusively with forces or powers invokes an infinite regress of interrelated dispositions that never engage with anything actual, and never amounts to a concrete reality. Nor is the continuity of macroscopic matter better understood in terms of a constitution of impenetrable Daltonian atoms, which are equally paradoxical. Collisions would be inelastic (otherwise, they have an internal structure of parts which would mean that they are not atoms), calling on infinite forces. Chemical combination, which amounts to more than mere juxtaposition of atoms, is without explanation, as are weaker interactions underlying differences of phase.

What there really is down there in very small regions for very short intervals of time is something of a mystery. Successively taking proper parts of a macroscopic body will not lead to ever smaller particles and ultimately, classical point masses. Moving from the macroscopic scale takes us to the mesoscopic scale of bodies ranging from micrometers down to bodies the size of collections of atoms such as make up a molecule. Macroscopic behaviour is distorted by statistical fluctuations and quantum effects begin to appear. At still smaller scales we are entirely within the quantum world, governed by indeterminacy relations forbidding simultaneous possession of exact momentum and position, and raising issues about the conjoint definiteness of time and energy (Hilgevoord 2002). We are told that classical particles emerge from the quantum world as a result of decoherence in which environmental effects nullify the interference in probabilities characteristic of quantum systems. With no pretensions to delve into this submicroscopic world of the really brief and very small, I prefer to leave incomplete the mereological theory of quantities, regions and times rather than making unreasonable claims about points, point-particles and instants or their atomic counterparts, and leave unanswered questions about how the boundary between the macroscopic and microscopic realms is to be articulated.

Chapter 2
Occupying Space

2.1 Boundaries

Aristotle's puzzle about starting and stopping was mentioned by way of initial motivation of the treatment of times as extended intervals without recognising the existence of instants as the boundaries of the intervals (Sect. 1.3). Since the system is made complete by adopting postulates about the infinite extent and denseness of time, it cannot easily be said to presuppose the existence of instants but merely not mention them. Analogous questions arise regarding the existence of boundaries in space, about which a similar position has been anticipated with the development of a theory of regions analogous to that of temporal intervals which is also complete when postulates determining whether space is finite or infinite in extent and discrete or dense are adopted. Underlying the adoption of these systems is the idea that the extended spaces and times are all that is needed to accommodate the spatio-temporal features of continuants and processes. These issues are pursued in greater detail—something that would be severely limited by merely adopting an understanding of parthood in a "basic spatio-temporal sense"—after considering how some philosophers have tried to motivate the introduction of boundary entities in dealing with the spatio-temporal features of continuants.

Varzi (2013, p. 11) quotes a suggestive passage of Leonardo da Vinci's:

> What is it . . . that divides the atmosphere from the water? It is necessary that there should be a common boundary which is neither air nor water but is without substance, because a body interposed between two bodies prevents their contact, and this does not happen in water with air. . . . Therefore a surface is the common boundary of two bodies which are not continuous, and does not form part of either one or the other, for if the surface formed part of it, it would have divisible bulk, whereas, however, it is not divisible and nothingness divides these bodies the one from the other. (Leonardo da Vinci, *Notebooks*, 1938, pp. 75–6)

Smith and Varzi (2000) adopt some ideas from this source with a little modification in defense of what they call their realist stance on the question of boundaries. They credit Brentano with the view that there is genuine contact between entities such as

P. Needham, *Macroscopic Metaphysics*, Synthese Library 390,
https://doi.org/10.1007/978-3-319-70999-4_2

John's head and (the rest of) his body, understood to mean that "there are ... two boundaries (one belonging to the head and one to the body) which share exactly the same location" (Smith and Varzi 2000, p. 406). This is an example of what they call a fiat boundary, which relates the adjacent bodies symmetrically. Each body has its own boundary, and these coincide. Two material bodies cannot coincide, but boundaries can because, in Leonardo's phrase, they are "not possessed of divisible bulk" (Smith and Varzi 2000, p. 416).

Fiat boundaries are contrasted with natural, or bona fide, boundaries, which are marked by a discontinuity as in Leonardo's example and mark an asymmetrical sense of adjacency. In cases of the latter sort, there is a single boundary belonging to one of the two adjacent bodies which is *closed*, whilst the other lacks a boundary and is *open*. As with Aristotle's example of starting and stopping, this seems to raise an obvious objection of arbitrariness in determining which body is closed and which open. But Smith and Varzi dismiss the difficulty and simply maintain that

> the open/closed distinction is not in and of itself counterintuitive. Indeed, in some cases it seems quite reasonable: ordinary material objects are naturally the owners of their boundaries (their surfaces, in effect), and there is nothing counterintuitive in the thought that the environments in which objects are embedded are open. (Smith and Varzi 2000, p. 407)

This is counterintuitive, relying on some tacit understanding that material on either side of a phase boundary (between, say, gas and liquid or between liquid and solid) can be distinguished as an object and its environment. They don't even attempt to justify this clearly arbitrary division and they certainly don't offer any general guidelines indicating how the distinction is to be applied. In Leonardo's example, the air is less dense than the water, which might be taken as reason to downgrade the air to the subsidiary status of environment. (Casati and Varzi (1999, p. 90) seem to think only solids are material, regarding gases as "immaterial, airy bodies".) But then a stone at the bottom of a glass of water is more dense than the water, and by this criterion would be the closed body with the water now constituting the open environment. They don't say that density is their criterion, but where we have a stone at the bottom of a glass of water in contact with the air, either the water is the environment with respect to one object (the stone) and not the other (the air), or both the stone and the air are counted environments of the water which for some reason never constitutes an environment, or some further (unstated) criterion determines when water is and when it is not the environment of adjacent bodies. It is ludicrous to suggest that any of these alternatives is intuitively plausible, let alone so obviously right as not to need justification.

If, as Smith and Varzi maintain, there is no need to introduce material bodies as boundary entities, the question remains whether the spatial and temporal ontology must include such entities in order to accommodate the relations between macroscopic bodies. The unsatisfactory arbitrariness of their manner of introducing such boundary entities provides some motivation for investigating whether we can do without them. I now turn to consider this suggestion.

2.2 Doing Without Boundaries

Consider the following passage:

> Descartes and Spinoza ... allow that the parts of any given body are separable one from another. Any given part can continue to exist without being adjacent to any other particular portion of matter. This is because the parts of the plenum that constitute bodies are moveable one from another (contrast the parts of Newtonian space, which are immobile). (Holden 2004, p. 13)

What is it for an object to be immobile? Doesn't it mean that the object occupies one and the same place all the time, and is not able to occupy another region at any time? But this is not the reason that space is not moveable because it doesn't fulfil the first conjunct, even if the second conjunct might, for this same reason, be construed as true. Space doesn't occupy space; a region doesn't ever occupy another region. Holden may take space to be immobile because it doesn't make sense to say that space moves. But it doesn't seem to be anything like this that is being denied by allowing that the parts of a body are moveable from one another. Rather, it is that they are not somehow irremediably stuck together. The suggested contrast is a superficial one that breaks down as soon as anciliary concepts, over and above the strictly mereological ones, are introduced explicitly to make the notion of movement on which it depends clear.

A body will be said to *move* (be *moveable*) if it occupies one region at one time and (possibly) another at another. Reserving the term "separate" for its strictly mereological sense introduced earlier, a body is said to be *distant from* another at a time if the regions occupied at that time by each are each parts of distinct strata between which there lies another stratum. A body is said to be *removed (removeable)* from another if they abut at one time and (might) not at another. This calls for the introduction of a new primitive relation of occupying relating individuals, regions and times, in terms of which a definition of a time-dependent relation of abutment between bodies can be formulated.

A continuant, x, is related to space and time by virtue of occupying a certain region, p, for a certain time, t, expressed here by writing $Occ(x, p, t)$. (Predicates which relate different kinds of entities—referred to by different kinds of variable—will be written in this manner, with brackets and commas, in contrast with the way betweenness, for example, is written as holding between three regions as $Bpqr$, without brackets or commas.) The occupies relation will be interpreted to mean that the continuant in question occupies exactly the region in question for the time in question, and so conforms to the axiom:

$$Occ(x, p, t) \wedge p \neq q \;.\supset\; \sim Occ(x, q, t).$$

An alternative course would be to take $Occ(x, p, t)$ to imply that x occupies all the subregions (proper parts) of p at t, and to work consistently with this interpretation instead. But the course adopted here will prove to be the most useful. The alternative concept is not lost; a body, x, is said to *cover* a region p during t, symbolised $Cov(x, p, t)$, iff $\exists q\,(p \subseteq q \wedge Occ(x, q, t))$.

A natural first shot at defining abutment for two individuals, x and y, is to say that they occupy abutting regions at t:

$$A(x, y, t) \;\equiv\; \exists p\, \exists q\, (Occ(x, p, t) \;\wedge\; Occ(y, q, t) \;\wedge\; Apq).$$

(Although the two relations of abutment are each symbolised by A, they are of different arity, and the style and number of the variables appearing after "A" together with the convention that the variables following a predicate are enclosed in brackets where they are of different sorts suffices to distinguish them.) A book lying on a table, for example, abuts the table as defined. This shows that it is possible to give an account of two bodies being in contact—as close as can be with no body interposed between them yet not actually interpenetrating—without resorting to a notion of a boundary. It suffices to deal with the sort of cases considered by Smith and Varzi, at least as they seem to understand them. France and Germany are not separated as a natural phase boundary separates heterogeneous parts of a mixture like the oily and watery parts of a vinaigrette dressing. But the present solution is just as applicable as that offered by the fiat boundaries of Smith and Varzi. Questions can be raised about how far down into the earth's core France and Germany extend, and how far into the atmosphere. On the present approach, it would have to be determined what the objects are that actually abut. But analogous issues arise for Smith and Varzi. Fiat country boundaries cannot be simple lines but must extend over a surface.

Other questions can be raised about what the actual bodies are that are at issue here. This is clearly a problem when considering the relation between John's head and the rest of his body when they are connected with a continually moving blood stream, which is taken up in Sect. 2.4. Putting this issue aside for the moment, assume the individuals in question comprise solid bodies that don't exchange matter with their environment. Still the above definition doesn't take into account all the ramifications of the time at issue being an interval. There is a possibility that one or both of two bodies might abut while expanding, contracting, or undergoing deformation during this time, or they might both undergo wholesale motion during this time, raising the prior issue of how "occupies" is to be interpreted in order to take these possibilities into account.

2.3 Keeping in Touch

The region occupied during a time might vary in the course of that time. This may be due to translational motion of the body, or internal movement such as vibration, contraction, deformation or expansion, all of which I take to be subsumed under the term "motion". An object in motion during a given time is naturally understood to occupy the region it sweeps out during that time. This will cover the special case in which it is stationary throughout the period, when the region "swept out" is one and the same region occupied at any subinterval of this time. The general principle at issue is here called a *principle of accumulation*, and can be conveniently expressed in the following manner:

$$Occ(x, p, t) \supset \forall q\, (\exists t' \subseteq t\, Occ(x, q, t') \supset q \subseteq p).$$

An alternative thought might be to consider the region occupied for a time to be the common part covered (forming a subregion of) every region occupied at any subinterval of the time in question. But for an object moving further than its own length in the time in question, there would be no such common part.

The second implication cannot be replaced by an equivalence; some subregions of p might be too large, small or scattered to be occupied by x at any time. A point of detail disguised by this generalisation is a continuity condition to the effect that

$$Occ(x, q_1, t_1) \wedge Occ(x, q_2, t_2) \wedge At_1t_2 \,.\!\supset\, Cq_1q_2,$$

where C is the relation of being connected (abutting or overlapping). The region p occupied throughout an interval is then the region swept out during the interval, being a sum of its subregions occupied by x during subintervals in the sense that

$$\begin{aligned} &Occ(x, p, t) \supset \forall t_1 \forall t_2\, (t = t_1 \cup t_2 \supset \\ &\qquad \exists q_1 \exists q_2\, (Occ(x, q_1, t_1) \wedge Occ(x, q_2, t_2) \wedge p = q_1 \cup q_2)). \end{aligned}$$

If general sums of spatial regions were allowed, this could be more compactly written as

$$Occ(x, p, t) \supset p = \Sigma q\, \exists t' \subseteq t\, Occ(x, q, t').$$

But in any case, since sums are unique, the principle of accumulation encompasses the point of interpretation already mentioned, that occupying a region means occupying exactly that region. Accordingly, bodies occupying distinct places at the same time are distinct, and a given body never occupies two places at the same time. In view of the uniqueness of the region occupied, the occupies relation $Occ(\pi, p, t)$ can be equivalently expressed as a function, $Occ(\pi, t) = p$, which will sometimes facilitate the formulation of a principle.

Once the possibility of movement is recognised, it can be seen that the definition of abutment proposed in the previous section doesn't even provide a sufficient condition. It would allow that x and y abut if they pass by one another during the time at issue without ever coming really close, but such that the regions swept out abut. The one might have missed the boat, as it were, because the other took off earlier.

We might try to remedy this by requiring abutment to hold throughout whenever it holds:

$$\widehat{A}(x, y, t) \equiv \forall t' \subseteq t\, \exists p\, \exists q\, (Occ(x, p, t') \wedge Occ(y, q, t') \wedge Apq).$$

But this won't quite do the trick because it doesn't allow for two moving bodies continually abutting while in motion and sweeping out regions that overlap. Although it is possible that when on the move, abutting bodies don't occupy overlapping regions, that depends on the shapes of the bodies and the direction of the motion in relation to the regions occupied immediately prior to the time in question. What has just been defined is adequate for the special case (recorded by the circumflex) where there is no overlapping of regions occupied. But in general, abutting bodies could well occupy regions which overlap (at least for part of the time at issue), and this possibility must be taken into account. The temptation is to say that they abut instantaneously throughout the interval. But since instants are not recognised here, the challenge is to address the issue without resorting to a conception of instantaneous abutment.

Where there is overlapping, the common region covered by two abutting bodies on the move is ever smaller for smaller subintervals of the time of abutment. But what is required is not just a matter of approaching an ever smaller region for ever shorter times, as the denseness of space and time would allow, since this could be an ever closer approach to a fixed region as times become indefinitely shorter. All subregions of the region common to those swept out during the time in question must be considered. The possibility that two individuals may spatially overlap although they cannot abut because one is a component part of the other, as the engine is a component part of a car, must also be taken into account since the region swept out by a car has less in common with that swept out by its engine the shorter the time. Using "$x \oplus_t y$" to symbolise the relation "x and y have a common component part during t" symmetric in x and y (discussed and defined in the next chapter (Sect. 3.4)),[1] abutment on the move can then be defined by:

$$\begin{aligned}\widetilde{A}(x, y, t) \;\equiv .\; & Occ(x, t) \circ Occ(y, t) \wedge {\sim}(x \oplus_t y) \wedge \\ & \forall p, q\, (\exists t', t'' \subset t\, (p = Occ(x, t') \cap Occ(y, t') \wedge \\ & q = Occ(x, t'') \cap Occ(y, t'') \;\wedge\; t' \subset t'') \;\supset\; p \subset q),\end{aligned}$$

where the functional form of the occupies relation has been used.

Drawing these proposals together in a general definition, it should be recognised that bodies abutting during a time may or may not occupy overlapping regions during one or more subintervals of this time. Where there is no overlapping of regions occupied, the above proposal for abutment between bodies under this condition (i.e. $\widehat{A}$) is adequate. But the alternative must be incorporated into a general definition, which replaces the definition of A given in the previous section:

$$A(x, y, t) \;\equiv\; \forall t' \subseteq t\, (\widehat{A}(x, y, t') \vee \widetilde{A}(x, y, t')).$$

The alternative covered by $\widehat{A}$ is not confined to the case of stationary or static abutment where there is no change of regions occupied, but it should be possible to

[1] A stronger requirement would be that the individuals have no constituent matter in common. But as we will see shortly (Sect. 2.4), abutting individuals may exchange matter.

characterise this special case. The conditions defining $\widehat{A}$ are necessary for abutment of stationary bodies, but not sufficient since two bodies might continually abut while in motion without any overlap of the regions they sweep out. Static abutment, covering the case when there is no translational motion or change of shape or volume, must imply that each body occupies one and the same region throughout the time in question. This can be expressed by moving the existential quantifiers in the definition of $\widehat{A}$ to the left of the universal quantifier, and defining static abutment, *SA*, by

$$SA(x, y, t) \equiv \exists p \exists q \forall t' \subseteq t\,(Occ(x, p, t') \wedge Occ(y, q, t') \wedge Apq).$$

This can hardly be described as an elegant solution to the problem of keeping in touch, but in as much as it does the job, a serious potential objection to the adequacy the pointless ontology of spaces and times is overcome.

2.4 Abutment Without Material Abutment

Does Leonardo's example of nothingness dividing the water from the air present us with a case of two connected material bodies abutting? That depends on what is meant by a material body. For only very short periods at best, the water and other material making up the sea abuts the nitrogen, oxygen, etc. making up the air. Modern science tells us that, as conditions of temperature and pressure vary, the proportion of water vapour in the air varies, as do the proportions of the various substances from the air—oxygen, nitrogen, carbon dioxide, etc.—dissolved in the water. Parts of the liquid phase are transformed into the vapour phase and enter the air, or parts of the water vapour in the air enter the liquid phase, along with parts of the oxygen, nitrogen, carbon dioxide, etc. in the air which dissolve in the liquid. Even when equilibrium obtains at constant temperature and pressure, there is still an exchange of microparticles between the phases. While equilibrium holds at the macroscopic level, there is a dynamic equilibrium at the microscopic level in which the more energetic water particles leaving the liquid phase are matched by less energetic water particles in the gas phase entering the liquid phase. Thus, a quantity of air immediately above a quantity of water, as in Leonardo's example, which is connected at some relatively short time will not remain so for long but become scattered as parts dissolve in the water; likewise for a quantity of water immediately below such a quantity of air. Consequently, these quantities, which may abut in the sense of occupying abutting regions for a very short period, will not do so for longish times on the macroscale, of the order of a microsecond (10^{-6} s.) and longer.

What is required to accommodate Leonardo's example to the modern understanding of matter is a distinction between the notion of an individual and its constitutive matter, the individual remaining the very same thing throughout the variations in its material constitution. The sea and the air are often thought of as individual bodies in this sense in everyday contexts, as for example when we talk of the air becoming more polluted and increasing its proportion of carbon dioxide. To seal the distinction with a name, the term "atmosphere" might be called on for this use, and the term "air" reserved as a predicate applicable to its constitutive matter (or at least the gaseous part). Similarly, "the sea" names the individual and the term "water" can be reserved as a predicate to describe some of its constitutive matter. Then we can say that at least some of the air dissolves in the sea, but not that the atmosphere does so, and that some of the water from the sea, but not the sea itself, enters the atmosphere. A more graphic example is Smith and Varzi's abutting head and body, which stand in this relation despite the continual circulation of blood between the two. We might well say that these are two abutting material bodies, but it is not the matter that abuts.

Individuals are related to quantities by being constituted of them, typically different quantities at different times. The constitutive matter of an individual doesn't necessarily vary over time. The marble or bronze of a well-preserved statue is to all intents and purposes constant throughout the life of the statue (whose demise is by no means the demise of its constitutive matter). But in general a quantity, π, stands to an individual, x, as its constitutive matter for some part of the lifetime of the individual, and so the constitution relation is a triadic one, $Const(x, \pi, t)$. An individual like the atmosphere exchanges matter with its environment pretty much continuously. Over aeons, we know that its chemical composition has changed entirely. Ammonia has disappeared and oxygen appeared as the result of chemical processes in matter on the earth's surface. Hydrogen has been lost to outer space because the gravitational pull couldn't hold it back.

In the special case where an individual's constitutive matter remains fixed, the region occupied by the individual is just that occupied by its matter. But where the constitutive matter varies, the situation is not so simple. Two ways of thinking about an individual's constitutive matter at an interval of time during which the matter varies suggest themselves. Either this matter is the mereological product of the constitutive matter at all subintervals of the time in question, i.e. what remains throughout this time, or it is the sum of all that is constitutive matter for any subinterval of the time in question. These alternatives coincide where the constitutive matter is fixed. But where this is not so, the product in the first option becomes ever smaller as the time in question becomes longer, and would not be defined for the atmosphere for a time stretching over aeons during which all the original matter ceases to constitute it and is completely replaced by other matter since mereology doesn't allow that there is a null quantity. On the second alternative, the constitutive matter satisfies an accumulation condition of the same general kind already introduced for the occupies relation,

$$Const(x, \pi, t) \supset \pi = \Sigma\rho\, \exists t' \subseteq t\, Const(x, \rho, t'),$$

and this is how the constitution relation will henceforth be understood.

Neither alternative offers a satisfactory way of reducing occupancy by an individual to occupancy by its constitutive matter, and I see no alternative which does. The problem is compounded by the fact that an individual might grow or diminish in size during the time in question. Accordingly, both of the predicates $Occ(\pi, p, t)$ and $Occ(x, p, t)$ are treated as primitive. Occupying by quantities is likewise interpreted as occupying exactly a given region at a time:

$$Occ(\pi, p, t) \wedge p \neq q \;.\!\supset {\sim}Occ(\pi, q, t),$$

and satisfying a spatial accumulation principle analogous to that satisfied by individuals:

$$Occ(\pi, p, t) \supset \forall q\, (\exists t' \subseteq t\, Occ(\pi, q, t') \supset q \subseteq p).$$

There is not in general an analogue of the continuity condition because a quantity might be scattered. But the region p occupied for a time by a quantity is a sum of its subregions occupied by the quantity during subintervals in the sense that

$$\begin{aligned} Occ(\pi, p, t) \supset \forall t_1 \forall t_2\, (t = t_1 \cup t_2 \supset \\ \exists q_1 \exists q_2\, (Occ(\pi, q_1, t_1) \wedge Occ(\pi, q_2, t_2) \wedge p = q_1 \cup q_2)). \end{aligned}$$

Sums of regions are related to sums of quantities via the further principle that the region occupied by a sum of quantities is the sum of the regions occupied by these quantities. Capitalising on the functional form of the occupies relation, $Occ(\pi, t) = p$, this principle can be neatly expressed in terms of the binary sum as follows:

$$Occ(\pi \cup \rho, t) = Occ(\pi, t) \cup Occ(\rho, t).$$

More generally, it would have been convenient to be able to speak of the region occupied by the sum, $\Sigma\pi\, \varphi(\pi)$, of quantities satisfying a condition $\varphi(\pi)$ by writing

$$Occ(\Sigma\pi\, \varphi(\pi), t) = \Sigma p\, \exists\pi\, (\varphi(\pi) \wedge Occ(\pi, t) = p).$$

But summation over regions was confined to the binary sum (and its iterations) in Sect. 1.4, and this mode of expression is not available.

Occupying by quantities and individuals is related by the principle

$$Occ(\pi, p, t) \wedge Occ(x, q, t) \wedge Const(x, \pi, t) \;.\!\supset q \subseteq p,$$

which puts a constraint on the region occupied by an individual by its constituent matter, but no more than that. Where an individual's constitutive matter does vary over time, it will in general occupy a more inclusive region than that occupied by the individual at the same time.

Again, since sums are unique, the principle of accumulation for the region occupied by a quantity entails that the principle that occupying a region means occupying exactly that region also applies to quantities. Accordingly, quantities occupying distinct places at the same time are distinct, and a given quantity never occupies two places at the same time. From this we deduce that, if static abutment, *SA*, of two quantities is defined as it was for individuals, by

$$SA(\pi, \rho, t) \equiv \exists p \, \exists q \, \forall t' \subseteq t \, (Occ(\pi, p, t') \wedge Occ(\rho, q, t') \wedge Apq),$$

then quantities occupying abutting regions at a time are separate. For every part of the one region is distinct from every part of the other and so occupied by distinct parts of the two bodies, which therefore have no part in common. The definition of static abutment for quantities thus implies that abutting quantities are separate, which speaks for the adequacy of the definition.

This explicates what is affirmed and denied in the claim that the atmosphere statically abuts the sea but not the constituent air and water (for the same time).

Finally, the atmosphere differs from many examples of individuals in one respect which is worth noting. Thanks to the distinction between an individual and its varying constitutive matter, the atmosphere can be said to abut the earth, i.e. the land and sea, for appreciable times during which the constitutive air and water don't abut. The atmosphere is bounded below, as it were, in the sense that it is in contact with (statically abuts) the land and the sea. But it is not clear that the atmosphere is bounded above, and that it exactly occupies a finite region. This is not just a matter of there being no quantity of matter or no individual abutting it from above. The density of matter gets indefinitely smaller moving away from the earth. Suitable infinite regions would be available if the sum operation on regions were not limited to binary sums but admitted general sum. As matters stand, however, with no general sums of regions, it seems that we have to recognise that some individuals do not occupy any region, although there are regions that they cover. The definition of static abutment given at the end of Sect. 2.3 must be modified if we still want to say that such individuals abut others, as the atmosphere abuts the earth. In such cases, we might appeal to the notion of a component part defined in the next chapter (Sect. 3.4), and allow that an individual statically abuts another if either it or one of its component parts does.

2.5 Mathematical Analysis and Indispensability

Material objects on the macroscopic scale are generally thought of as continuous. This is not to deny that there are heterogeneous quantities of matter, but that sufficiently small quantities are homogeneous while still on the macroscopic scale. In other words, a heterogeneous quantity has homogeneous parts. Macroscopic theory that systematically treats such objects is typically formulated in terms of the differential and integral calculus of classical mathematical analysis. This captures the idea of a smooth function that moves along successive values as the argument successively covers all the numbers within given limits which is not merely continuous, but also without kinks, that is, *differentiable* at all points in the range of argument values. Classical thermodynamics already treats the energy and entropy as differentiable functions of the amounts of the various substances making up the mass of the body to which the theory is applied, which is certainly not to treat the arguments as ranging over integers as the discrete structure of matter would seem to require. Its extension to irreversible or non-equilibrium thermodynamics, where the thermodynamic magnitudes are treated as functions of a position vector and a time, is more grist to the mill. It also presents a challenge to the pointless ontology of space and time presented here. A literal interpretation of the theory would suggest that there should be objects with features corresponding to those ascribed by the theory to numbers in the infinitesimal limit invoked by differentiability. Nominalists have tried to show that we can dispense with the mathematical entities themselves. That is not the present issue, however, which is concerned with whether the theory should be literally interpreted with physical elements *corresponding* to the mathematical elements in the formalism.

A body is in a definite state when it is at equilibrium. Only then does a reading registered by a thermometer record a temperature and a reading registered by a pressure meter record a pressure. These are intensive properties, exhibited by the body when each of its parts exhibit the same temperature or pressure. Classical thermodynamics provides a theory within which the modern concept of absolute temperature is developed, independent of the properties of any particular thermometric material on which the traditional mercury, alcohol and gas thermometers are based, and with an absolute zero, colder than which it is impossible to become.[2] It is extended to a theory that copes with the effects of smooth differences in temperature, exemplified by a temperature gradient, throughout a body in the theory of non-equilibrium thermodynamics along the following lines.

In what Callen (1985) calls the entropy representation of a body treated by classical thermodynamics, the entropy is taken to be a function, $S(U, V, N_1, \ldots, N_k)$, of the other extensive variables, here taken to be the energy, U, and the volume, V, of the body comprising k different substances in amounts measured by N_i, $1 \leq i \leq k$. The condition of equilibrium is that the rates of change of the entropy with respect to each of the extensive magnitudes on which it depends is zero. These rates of change are intensive magnitudes of the entropy representation. $\partial S/\partial U$, for example, is the inverse of the temperature, $1/T$. If there are differences in the inverse temperature (because there are differences in the temperature) within the system—for example if there is a temperature gradient over the system—then this gradient acts as a kind of force, driving a heating of (a flow of energy into) one part of

[2]The negative absolute temperatures first obtained by Purcell and Pound (1951) do not infringe this principle; they are *warmer* than any positive temperature (see Zemansky and Dittman 1981, Ch. 19). The negative sign arises because temperature is equated with a derivative, which can be negative even though it is usually positive—see below.

the system by another until the gradient is eliminated and equilibrium established. When accommodating the formalism to deal with irreversible processes arising in non-equilibrium situations that drive the system towards equilibrium, it is assumed that the extensive variables on which S depends are functions of position, $\mathbf{x}$, and time t. Associated with the energy, U, then, are three components along each of the Cartesian coordinates and their corresponding rates of change, called the energy fluxes in each direction, which are the components of the vector, $\mathbf{J}_U$, for the energy current density giving the amount of energy flowing across unit area in unit time. Similarly, there are fluxes of each of the different substances describing the flow of a particular substance per unit area per unit time.

Entropy is defined in classical thermodynamics for bodies at equilibrium. The concept is adapted to non-equlibrium situations by postulating that

> To an infinitesimal region we associate a local entropy $S(X_0, X_1, \ldots)$, where, by definition, the functional dependence of S on the local extensive parameters $X_0, X_1, \ldots$ is taken to be identical to the dependence in equilibrium. That is, we merely adopt the equilibrium fundamental equation to associate a local entropy with the local parameters (Callen 1985, p. 310).

Equilibrium is thus assumed to obtain at sufficiently (infinitesimally) small locations, and the differential $dS = \Sigma_i F_i X_i$ where the F_i are the partial derivatives, $\partial S/\partial X_i$, of S with respect to the extensive variable X_i. Taking the magnitudes per unit volume—replacing entropy by entropy density, denoted by the corresponding small letter, etc.—this becomes $ds = \Sigma_i F_i x_i$, where the term for the volume, V, is removed from the summation.

A short argument leads Callen to the following expression for the rate of local production of entropy, $\dot{s}$:

$$\dot{s} = \sum_i \nabla F_i \cdot \mathbf{J}_i.$$

The gradient of the entropy-representation intensive parameter, ∇F_i, is the generalised force or *affinity* that drives the flux of the corresponding magnitude (energy density, concentration of kth substance, ...) to change in the direction of equilibrium. The affinity associated with the energy current density, for example, is $\nabla(1/T)$, and the affinity associated with the amount of the kth substance flowing through unit area in unit time is $-\nabla(\mu_k/T)$ where μ_k is the chemical potential of the kth substance. The theory of non-equilibrium thermodynamics developed on this basis deals successfully with irreversible processes arising in situations not too far from equilibrium in what is called the linear regime.

Without pursuing the details any further, this brief review indicates how assumptions and techniques of analysis are applied in setting up a theory of smooth variation of macroscopic properties over space and time. Focusing on one property in particular, the temperature, this is clearly a step in the direction of acknowledging that the concept can be extended to situations in which it varies smoothly over space and time. In this connection, Callen notes that

> *the local intensive parameter F_i is taken to be the same function of the local extensive parameters as it would be in equilibrium.* It is because of this convention, incidentally, that we can speak of the temperature varying continuously in a bar, despite the fact that thermostatics implies the existence of temperature only in equilibrium systems. (Callen 1985, p. 310)

Consider what this means if its formal representation is taken literally. A smooth function is not merely continuous in the mathematical sense, but differentiable at all points (which excludes angular chinks). A derivative is defined as the limit of a converging sequence of quotients of numbers, and such limits must exist everywhere in the range of the argument of a function if the function is to be differentiable everywhere. The rational numbers are dense—between any two there is a third—so that there are infinitely many rational numbers between any two. But that is not enough to ensure differentiability. There are gaps into which yet more numbers must be pressed to guarantee the existence of these limits. The real numbers are complete in the sense that the limit of every Cauchy sequence $\{x_i\}$, for which

$$\forall \epsilon\, \exists n\, \forall m\, (m > n \supset \mid x_m - x_n \mid < \epsilon),$$

exists. This is not the case with the rational numbers, which are not order-complete. It is difficult to see how a temperature gradient could be literally understood in this way.[3] The temperature would fall off at a constant or varying rate along a given Cartesian axis, so that the derivative of the temperature as a function of the Cartesian coordinate is defined with a definite value at each point. But at a sufficiently small scale matter is granular, and for a gas that is not too dense nor at a temperature close to a phase change, the temperature is determined by statistical features of the gas molecules that are not defined for sufficiently small regions containing only a few molecules, not to mention even smaller regions not large enough to be occupied by a molecule. The denseness of the rationals seems already to go beyond corresponding physical features in the sense that there isn't anything occupying points in space with a structure corresponding to that of triples of rational numbers at points in time similarly corresponding to the rationals which can bear a temperature. There certainly seems to be nothing corresponding to the limits of Cauchy sequences. It seems that the representation of smoothness at the macroscopic level by differentiable functions calls for features on a very small scale—below that of the scale of the microscopic constituents of macroscopic bodies—that cannot be interpreted realistically. What should we say about the Quinean doctrine that we can't shun the ontological commitments of our theories?

To repeat what was said earlier, there is no suggestion of nominalism here—i.e. the doctrine that denies that "numbers, functions, sets or any similar entities exist" (Field 1980, p. 1). I am denying that there is anything *corresponding to*

[3]The same applies to an alternative approach to the foundations of analysis that treats infinitesimals literally as entities supplementing the real numbers rather than eliminating them contextually in terms of limits (Bell 2008).

certain mathematical structures in the ontology of macroscopic objects with which the theory of irreversible thermodynamics deals.

A non-literal interpretation is described in Kondepudi and Prigogine's textbook:

> temperature is well defined when the velocity distribution is "Maxwellian" [when the probability that a molecule has a given velocity is specified by the Maxwell distribution].... In practice, only under very extreme conditions do we find significant deviations from the Maxwell distribution. Any initial distribution of velocities quickly becomes Maxwellian due to molecular collisions. Computer simulations of molecular dynamics have revealed that the Maxwell distribution is reached in less than 10 times the average time between collisions, which in a gas at a pressure of 1 at, is about 10^{-8}s. Consequently, physical processes that perturb the system significantly from the Maxwell distribution have to be very rapid. (Kondepudi and Prigogine 1998, p. 334)

They go on to illustrate what this means for the time when chemical reactions are at issue. Only a very small percentage of molecular collisions are *reactive collisions* in the sense that they lead to chemical change. Rates of an order corresponding to nearly every one of, for example, the 10^{31} collisions per litre per second in a gas at a pressure of 1 atm are extremely rare.

> Most of the reaction rates we encounter indicate that reactive collision rates are several orders of magnitude smaller than overall collision rates. Between reactive collisions the system quickly relaxes to equilibrium, redistributing the change in energy due to the chemical reaction. In other words, any perturbation of the Maxwell distribution due to a chemical reaction quickly relaxes back to the Maxwellian with a slightly different temperature. Hence, *on the timescale of chemical reactions, temperature is locally well defined.* (loc. cit., my emphasis)

A similar tale is told about the spatial dependency of the thermodynamic magnitudes. These magnitudes undergo fluctuations about their mean values, and a value for a thermodynamic magnitude can only be meaningfully associated with a small volume, ΔV, when the size of the fluctuations is very small compared with the value of the thermodynamic magnitude itself. "Clearly, if ΔV is too small, this condition will not be satisfied" (Kondepudi and Prigogine 1998, p. 335). They show that

> if $\tilde{N}$ is the number of particles in the considered volume, then ... the fluctuations $\delta\tilde{N} = \tilde{N}^{1/2}$. As an example, let us consider an ideal gas for which $N = \tilde{N}/N_A = (p/RT)\Delta V$. For a given ΔV it is easy to compute the relative value of the fluctuation $\delta\tilde{N} = \tilde{N}^{1/2}$. To understand how small ΔV can be, we consider a gas at a pressure $p = 1$ atm and $T = 298$ K, and compute the fluctuations in the number of particles $\tilde{N}$ in a volume $\Delta V = (1\mu m)^3 = 10^{-15}L$. We find that $\delta\tilde{N}/\tilde{N} \approx 4 \times 10^{-7}$. For liquids and solids the same value of $\delta\tilde{N}/\tilde{N}$ will correspond to an even smaller volume. Hence, *it is meaningful to assign a mole number density to a volume with a characteristic size of a micrometer. The same is generally true of other thermodynamic variables.* If we are to assign a number density to a volume ΔV, then the number density in this volume should be nearly uniform. This means that the variation of number density with position on the scale of a micrometer should be very nearly uniform, a condition satisfied by most macroscopic systems. This shows that a theory based on local equilibrium is applicable to a wide range of macroscopic systems. (loc. cit.)

Clearly, these authors defend the principles of continuity embodied in the theory on a sufficiently large scale of time and space. There is no question of taking the theory literally on a smaller scale, and certainly not of treating matter as comprising point particles occupying points of space corresponding to triples of real numbers at instants of time corresponding to real numbers. I venture to say that viewing the treatment of the properties of matter with the aid of the tools of mathematical analysis along these lines is standard practice for scientists.

Avoiding a literal interpretation of a theory is often taken as the hallmark of an antirealist stance. But ascribing such a view to Kondepudi and Prigogine is not easily reconciled with their clear affirmation of the reality of microscopic particles and the adequacy (substantial truth) of the macroscopic theory of non-equilibrium thermodynamics. It is misleading to put too much emphasis on the purely mathematical aspects of the theory. Specific empirical claims are routinely made precise by accompanying them with detailed specification of how the recognised sources of systematic error are accommodated and a proper statistical assessment of the probable limits of error due to the spread in the measurements. (As Duhem (1954, p. 169–71) makes clear, empirical claims lacking such qualifications are hopelessly vague, indicating no more than ignorance of how to specify likely error.) Much the same idea carries over to the manner in which scientists claim theories are true. Kondepudi and Prigogine circumscribe the intended domain of application of non-equilibrium thermodynamics in some detail, including limitations on the smallness of bodies as we have just seen. This is how they formulate their claim about the truth of the theory. There is no question of adopting an antirealist stance.

Kondepudi and Prigogine's qualifications might be taken to suggest an eliminative view of macroscopic reality, to the general effect that all there is is the microscopic realm. But as Kondepudi and Prigogine point out, the Maxwellian distribution gives the probability that a molecule has a given velocity as a function of temperature. Derivations of one or other form of the Maxwell velocity distribution rely on an identification of a term in statistical mechanics (usually denoted β) with $-1/kT$, where T is the absolute temperature introduced in classical equilibrium thermodynamics, by consideration of principles taken from thermodynamics (Zemansky and Dittman 1981, p. 284; Denbigh 1981, p. 352). The statistical features of the microscopic realm are developed in standard texts *in conjunction with* principles of classical equilibrium thermodynamics. There is no attempt to present the theory as a reduction of the latter to the former, and it behoves the would-be reductionist to show how such a reconstrual is possible without blatant circularity. It therefore seems that, just as the Quinean indispensability argument speaks for the existence of purely mathematical entities such as real numbers and functions, it also speaks for the indispensability of macroscopic theory, in particular classical equilibrium thermodynamics, along with the corresponding macroscopic ontology in providing a picture of the world that includes the microscopic realm. This is not a conclusive argument; it allows that dispensability might actually be demonstrated. But it is quite clear where the onus of proof lies. Short of actually rising to the challenge and giving us the demonstration, the would-be reductionist doesn't have any case at all.

2.6 Occupying a Region Exactly

How plausible is the interpretation of the "occupies" relation as occupying *exactly* a certain region at a time? In Sects. 2.2 and 2.4, an object o was said to occupy a region p exactly at t in accordance with the principle

$$Occ(o,p,t) \wedge p \neq q \ .\supset \ {\sim}Occ(o,q,t),$$

where o is either an individual or a quantity. There is no problem with wholesale movement or any vibration or rotation during t, which is accommodated by the principle of accumulation. But this notion of exactness may seem problematic in the light of the principle of infinite divisibility of space, which is included in the complete theory of spatial regions presented in Sect. 1.4. Probing the structure of a macroscopic object down to the molecular level, it might be difficult to say how the electronic structure of the outlying molecular structure of a physical object such as a table precisely delimits the region occupied by the table. Just how exactly can a body occupy a region?

The question is whether, at the level of electrons, any region of space is delimited at all. Electrons aren't the sort of thing that occupies regions. They are described by a wavefunction assigning a complex number to points in space at instants of time which when multiplied by its complex conjugate and integrated over a region yields the probability of finding an electron within that region.[4] Chemists look at the presence of electrons in molecules in terms of the variation of electron probability density over molecular structure. For many practical purposes, the distribution of 95% of the electron density usually covers a region not too far from the nuclei and is taken to provide a fair picture of the molecule. But there is no bound beyond which the electron density falls to zero. Consideration of the quantum behaviour of a table's microconstituents doesn't show the macroscopic region it occupies to be vaguely drawn; it fails to delimit any specific region within the whole of space as that within which its constituents are definitely located.

Occupying a region is a macroscopic concept applicable to macroscopic objects. Quantities are treated in science as having a specific volume, made precise by considering the statistical fluctuations along the lines indicated in the last section. Individuals that occupy a definite region may raise different issues—for example, how much of the air in the vicinity of a mouth and nose is part of the constitutive matter of a human being at a given time may be unclear. But this concerns the constitution relation, not the region occupied by specific quantities of matter or the region occupied by specific individuals, which is not defined by the region occupied by its constitutive matter. Considered in the abstract, it seems that space is infinitely divisible, as expressed by axiom MS9 of Sect. 1.4. Does the macroscopic

[4]The probability interpretation requires that the wavefunction is normalised, so that the total probability of finding an electron somewhere is 1. For this purpose the product of the wavefunction with its complex conjugate is integrated over *the whole of space*.

character of quantities require that this axiom be abandoned? It wouldn't be abandoned in favour of an axiom introducing spatial atoms—regions with no proper parts—which is not motivated by the considerations above. Perhaps the appropriate moral to draw is that what seems to be a reasonable goal of attaining a complete system for spatial regions when considered in the abstract doesn't necessarily carry over for a more comprehensive treatment dealing with objects that are not purely spatial and their relations to regions. Such a theory doesn't give a complete specification of the spatial parthood relation. What might seem unmotivated when considering the theory of spatial regions *in abstracto*, as a mathematical exercise in pure geometry, need not be so when part of a theory of material objects. The Holden message is that the purely mathematical interpretation of the properties of material objects is quite mistaken. How to make a positive suggestion in the light of this insight is less easily addressed.

How might the principle of exact occupation be softened to accommodate sensibly indistinguishable regions? Taking a cue from the treatment of infinitesimals by Bell (2008), we might toy with the idea of associating a region p with a *neighbourhood of indistinguishability*, Λ_p, such that for any indistinguishable variation, ε, in p, $p \cup \varepsilon$ and $p - \varepsilon$ are indistinguishable from p. (Following the analogy with Bell's infinitesimal neighbourhood of 0 (2008, pp. 2–6), members of Λ_p would not be distinct from p—i.e. not non-identical with p—without being identical with p.) The principle of occupying exactly a particular region would give way to a principle of the kind

$$Occ(\pi, p, t) \wedge Occ(\pi, q, t) \;.\supset\; I(p, q),$$

where $I(p, q)$ means that q is indistinguishable from p.

The problem with any such idea is that there is a fundamental point of disanalogy. Bell emphasises that the characteristic property of infinitesimals is that they are so small that their higher powers can be neglected and "the property of being a nilsquare infinitesimal is an *intrinsic* property, that is, in no way dependent on comparison with other magnitudes or numbers" (2008, p. 2). But the lower bounds of insensitivity to fluctuations depend on environmental circumstances such as temperature and pressure and the susceptibility of the particular kind of matter occupying a given region at a given time to the determination of its properties by these conditions, and are in no way intrinsic to particular regions or their occupants. In view of this, the most reasonable course would seem to be to uphold the infinite divisibility of space and time, and to understand the principle of exact occupation in the spirit of Kondepudi and Prigogine's remarks on making precise the variations in volume tolerated by macroscopic description. At all events, the infinite divisibility of space doesn't entail a corresponding principle regarding matter along the lines of

$$Occ(\pi, p, t) \wedge q \subset p \;.\supset\; \exists \rho \subset \pi\, Occ(\rho, q, t).$$

No such principle is affirmed here, and nor is any kind of opposing atomistic principle. The matter is left open.

Chapter 3
Constitution

3.1 Distinguishing Constitution and Parthood

Advocates, but more especially critics, of an ontology of continuants have saddled it with a notion of parthood that is time dependent on the strength of the supposition that one and the same continuant gains and loses parts over time. Sider claims that on the three-dimensional theory, "parthood-at-*t* must be primitive" (2001, p. 57). But a corresponding time-dependent operation of summation (if the proposal is to systematically reformulate mereology by replacing a primitive dyadic relation with a triadic relation) is not, it seems, faithful to the notion of a continuant. A sum is a unique collection, not one that can gain and lose parts. In fact the idea seems incoherent since the sum operation acts on entities of a given kind to yield an entity of the same kind, whereas a putative time-dependent sum of continuants doesn't seem to be a continuant. As we have seen, the standard notion of a sum does apply to quantities, yielding a quantity as an operation should. But quantities neither gain nor lose parts over time. Continuants that gain and lose material are individuals, and these are not sums of their material or of anything else.

Sider seems to suggest that ordinary mereology (with the standard time-independent, dyadic relations and operations) must deal with four-dimensional objects. But quantities are just as much continuants as are individuals, and the fact that they sustain ordinary mereological relations and operations no more implies that quantities are four-dimensional than do the mereological relations and operations sustained by times and spatial regions imply that they are four-dimensional objects. Quantities' relation to time can be seen from the time-dependent relations they sustain, exemplified by the substance and phase properties mentioned shortly and explored in greater detail in later chapters, as well as the constitution and occupation relations discussed in some detail below. In the interests of clarity, where distinctive applications of standard mereology in the sense of an algebraic system of dyadic relations and operations are present, it will be prudent to reserve the terminology of standard mereology for mereological concepts and find other terms for concepts

P. Needham, *Macroscopic Metaphysics*, Synthese Library 390,
https://doi.org/10.1007/978-3-319-70999-4_3

which, despite any superficial linguistic analogies, are not mereological ones. The constitution relation, relating individuals to their accruing and discarded matter, though superficially analogous to parthood, is a case in point.

The beginnings of a distinction amongst material objects, between individuals and the quantities of matter of which they are constituted, appeared in the last chapter. The atmosphere, which abuts the sea, is not the same as the mixture of gases we call the air, which doesn't abut the saline solution constituting the sea, and the abutting of a head with its body doesn't imply that their constitutive matter abuts. The blood in particular flows continuously between heart and head. Constitution is a triadic relation between an individual, a quantity of matter and a time, the principal features of which are developed in this chapter. Other aspects of individuals that some philosophers have sought to capture by relativising parthood to time will be addressed by calling on the mereological features of those entities standing in the constitution relation that do lend themselves to treatment with standard mereology, and the difficulties of developing a sum *operation* on individuals will be discussed in greater detail.

The general idea of distinguishing individuals from their constituent matter is by no means unfamiliar, although terminology varies. But the viability of the notion isn't a matter of linguistic distinctions. Wiggins (2004, p. 36) suggests there is a fourth sense of "is", in addition to the copula, existence and identity, namely the "is" of constitution, but his example, "The soufflé you are eating is flour, eggs and milk", is unconvincing. As Pickel (2010, p. 7) says, a more natural way of putting this would be "The soufflé you are eating is made from flour, eggs and milk". The preposition would seem to be all-important in distinguishing between the original ingredients that the souflé was made *from* and what is there after mixing and heating, which the souflé is made *of*, the latter being a direct expression of the constitution relation. A better illustration of Wiggins's proposal would be "Humans are (mostly) water".[1]

Use of the "made from" idiom suggests that some corresponding use of the "made of" idiom would be appropriate if only we knew enough about the chemistry of the "making from" process. I've no idea what a cooked soufflé is made of, but I don't doubt there is something. This applies to the general idea that Arthur Prior nicely formulated in the following passage:

> ... the things of which we make our predications ... include things that have not always existed and/or will not always do so ... [These] countable 'things' are made or grow from bits of stuff, or from other countable 'things', that are already there. (Prior 1967, p. 174)

The plausible picture of middle-sized ontology sketched here portrays individuals ("countable things") as constituted of matter ("stuff") or comprising other individuals. It is not susceptible to the general criticism that constitution is identity, which

[1]Almotahari (2013) makes a linguistic case against Kit Fine's argument for distinguishing an object from its constitutive matter based on claims such as "The statue is Romanesque but the piece of alloy of which it is made isn't". But this argument plays no role in justifying the distinction as advanced here.

can't apply to a theory with the ambition of capturing Prior's "made of" relation because this is clearly a time-dependent relation, i.e. a three-place relation between an individual, a quantity of matter and a time. Whereas individuals are transient beings, existing for some time and not at other times, quantities "are already there" as Prior puts it, i.e. they are permanent existents. Consequently, an individual is constituted of a quantity of matter for no longer than the time it exists. In fact many individuals, including but by no means exhausted by biological organisms, exchange matter with the environment, and a particular quantity may constitute an individual for a considerably shorter time than the individual's lifetime. This conception of constitution will be elaborated in sufficient detail to make it clear that the objection fails. The mereological accounts of quantities, regions and times shows that they have entirely different kinds of identity conditions from those of the individuals constituted by quantities at times.

Quantities conform to the fundamental principle of the non-destruction of matter employed in chemists' understanding of chemical reactions at least since Lavoisier "demonstrated", as Poincaré (1913, p. 65) put it, "the indestructibility of matter by demonstrating the invariability of mass". In accordance with what is represented by the familiar balancing equations representing chemical reactions, quantities of matter bearing properties of being such-and-such substance kinds are brought into contact and undergo reaction, when parts combine with other parts resulting in these very same quantities bearing new substance properties. The mereological sum of the quantities at the end of the reaction is identical with the mereological sum of the quantities introduced at the start of the reaction. Change in chemical kinds is a matter of quantities loosing and acquiring different substance properties, which are relational, expressed by two-place predicates applying to a quantity and a time. Electrolysis of water under standard temperature and pressure, for example, transforms a quantity so that it no longer possesses the features of being water, liquid and spatially connected, acquiring instead the features of being partly free hydrogen, partly free oxygen, and entirely gas distributed throughout the atmosphere. Quantities' transtemporal identity conditions are simply their identity conditions—same parts, same thing. This shouldn't be confused with conditions of preservation of properties such as being a particular substance kind or in a specific state of mixture such as being a solution or heterogeneous matter exhibiting three phases, which depend on the appropriate criteria for the sustaining of these features over a period of time drawn from chemical theory.

Quantities of matter constituting familiar objects have undergone many chemical changes over the billions of years during which their elemental components were formed in the course of the evolution of stars and eventually found their way to our corner of the universe. The elements have combined to form water, carbon dioxide, minerals, carbohydrates, proteins and all the other compounds which, in various kinds of mixture, make up the atmosphere, the oceans, geological structures and biological organisms. The chemical substance kinds born by quantities for particular intervals of time are important factors determining whether they can constitute particular individuals at these times. But what chemists call the phase is also important. The water forming part of the constitutive matter of the sea and human beings must be in the liquid phase, whereas the water forming part of the constitutive matter of the Antarctic ice cap and the comet Hale-Bopp is in the solid phase. Tectonic plates would not give rise to earthquakes at their boundaries unless their constitutive matter was solid, and their material wouldn't be renewed as it is unless it could change from solid to liquid and from liquid to solid under varying conditions of temperature and pressure.

These features of quantities are pursued in greater detail in ensuing chapters. This outline should be sufficient for the purposes of understanding the features ascribed to individuals and the constitution relation in the following sections.

As these last few examples illustrate, many properties assigned to individuals are in fact, or directly derived from, properties of their constitutive matter. The melting point of water (or more generally, the phase boundary separating solid and liquid as pressure varies) is the property of the stuff water systematically recorded in thermodynamic studies which allows Hale-Bopp and the Antarctic ice cap to have a shape which is not determined by a solid container or imposed forces (like the shapes of liquids and gases) and which is relatively stable under the influence of ambient forces. This understanding of the bearing of general features of kinds of matter on the features of individuals is directly opposed to the suggestion (Robinson 2004, p. 21) that distinguishing the clay constituting a statue from the statue would give us two objects each weighing, say, 10 pounds which should have a total weight of 20, and not just 10, pounds. This is quite clearly gratuitous; the statue's weight is the weight of its constitutive matter, which shouldn't be counted twice. One and the same individual weighs 10 kg when a baby and 70 kg as an adult because of its changing constitutive matter, not 20 and 80 kg, or 80 and 140 kg. Robinson obviously has a different view, but it goes no deeper than the bare statement of it. No argument is offered in support of it which an opponent could address. Sider's (2001, p. 141) criticism of the idea that an individual is constituted of a distinct entity on the grounds of impenetrability fares no better. He simply assumes without offering any justification or overview that the laws of physics give him this conclusion (see Sect. 3.3 below).

Critics such as Robinson's three-dimensionalist and Sider, a four-dimensionalist, may each find that objects in their own preferred ontologies make no call on a constitution relation distinct from identity, but that doesn't amount to an objection to the present proposals. Sider chooses to criticise a constitution theorist who maintains that individuals coincide with their constituent matter in the sense of having parts all of which are common (Sider 2001, pp. 141–2). But this is just to define coincidence as identity by the mereological criterion, and simply assumes without arguing from generally accepted premises that constitution is identity. It recalls Sider's own remark that many of the arguments against four-dimensional entities are irrelevant or plain silly, often illicitly treating them as three-dimensional entities (such as Judith Jarvis Thomson's claim that stages constantly come into existence *ex nihilo*). Individuals as understood here don't coincide in Sider's sense with their constituent matter because they don't have any parts in common with their constituent matter. A more relevant concern is the relation between the regions occupied by an individual and its constitutive matter, which is taken up in Sect. 3.3.

3.2 Individuals

The changing character of the material content of identifiable objects is the major reason for introducing a constitution relation and distinguishing individuals from their constituent matter. With a primitive three-place relation of being constituted of standing between objects, quantities and times, individuals are taken to be objects that stand in the constitution relation to some quantity at some time. I see no reason why this shouldn't include the special case of individuals which have a fixed constitution and are constituted of the same quantity whenever they are constituted of any quantity of matter, leaving it open that some individuals merely might be, although they actually aren't, constituted of different quantities of matter. But the modal argument for distinguishing individuals and their constituent matter won't play any role in the present discussion. Note, however, that the case envisaged by Hawley (2004, p. 7) of a statue and a large lump of clay being formed simultaneously and subsequently destroyed simultaneously, misconstrues the notion of fixed constitution as the constitution relation is understood here. Again, I see no reason why individuals shouldn't include artifacts, although I do not rely on appeal to artifacts. What constitutes a statue, however, is a quantity of matter, which, even if it could exhibit the chemical properties of clay and the physical features of a lump for precisely the lifetime of the statue, would be neither created when the statue is nor destroyed when the statue is.

As understood here, an individual is constituted of exactly one quantity at a given time. This allows us to talk of parts of the quantity partially constituting an individual at a given time if they are parts of the quantity constituting the individual at that time. A question arises about how this unique quantity is to be understood analogous to that which arose for the occupation relation. Consider an interval of time t, not necessarily the entire lifetime of the atmosphere, but a considerably shorter one, say a day, a minute or a second. The chemical process of photosynthesis will be taking place wherever green plants are illuminated by the sun, extracting carbon dioxide from the atmosphere and combining it with water to produce carbohydrates. At the same time animals will be removing oxygen from the atmosphere as part of their metabolism, generating carbon dioxide which is put into the atmosphere. These and many other chemical processes together with physical processes such as evaporation and condensation contribute to the continual interchange of matter between the atmosphere and things on the earth's surface. What, then, is the constitutive matter of the atmosphere at time t? We might consider narrower and broader sums as possible candidates. On the one hand, there is the narrower mereological sum of all quantities of matter occupying some part of the region forming a thin shell around the earth throughout the whole of t. On the other hand, there is the mereological sum of all quantities of matter occupying some part of the region at some time during t. The former sum excludes all matter entering the atmosphere after the beginning of the period in question as well as all matter leaving during this time. Over sufficiently short times, the defining condition will be satisfied and the sum actually exist. It is part of the second sum, which also includes the matter excluded from the first sum, and the two sums converge as the time in question is taken to be shorter.

Either of the broader or the narrower sum would satisfy the requirement that the constituting quantity is unique since mereological sums are unique, but there is a possibility that, for sufficiently long times, the narrower sum doesn't exist because there is no null quantity. On the grounds that an individual is always constituted of matter, the constituting quantity is taken to be the broader sum and the following *principle of accumulation* adopted:

(PA) For any time *t*, if a quantity constitutes an individual at *t*, then that quantity is the mereological sum of all quantities constituting that individual for some subinterval of *t*.

Formally,

$$Const(x, \pi, t) \supset \pi = \Sigma \rho \exists t' \subseteq t\, Const(x, \rho, t'),$$

where $Const(x, \pi, t)$ stands for "x is constituted of π at t". Accordingly, the constitution relation is not in general to be construed to the effect that an individual is constituted of a quantity of matter throughout (at all subintervals of) some time *t*. Further, a proper part of a quantity constituting an individual at *t* doesn't constitute that individual at *t*. Should the need arise, such a proper part can, as already noted, be said to partially constitute the individual at *t*, although it must be emphasised that the mereological relation obtains between the two quantities. Constitution is not a mereological relation between a quantity and an individual.

The principle of accumulation entails that the weight of an individual as recorded by some such method as weighing on a bathroom scale is not in general the weight of its constitutive matter over a period longer than the weighing process. The latter, assuming it can be suitably assembled and the gravitational force on its mass measured, will in general increase with the length of time that the matter is the individual's constitutive matter, whereas the weight of an individual shows nothing like the same variation. Suppose I determine my weight on a bathroom scale to be 75.0 kg. If I repeat the measurement at regular intervals over a longish time (a day or more), or even remain standing on the scale, assuming it monitors continuously (and assuming that I remain perfectly still so as not to create spurious forces), my weight remains constant (within the limits of sensitivity of the bathroom scale), or if there is any variation, it doesn't show the same pattern of variation as does my constitutive matter over the same day (or more). There is no question of counting the same weight twice (cf. Robinson's objection mentioned in the last section). For the very short interval which is the time the bathroom scale takes to complete the process of recording my weight, my weight is the weight of my constitutive material, but not for longer times. If the bathroom scale were far more sensitive than such scales typically are, distinguishing the weights of masses to say $\pm 10^{-10}$ kg, with a much shorter response time for the measurement process, then the time when the individual is ascribed a weight would have to be correspondingly narrowed down much more finely than as it is ordinarily vaguely understood as the weight of a five-year old child, a middle-aged man, after the Christmas festivities, and so on. (And other factors bearing on the systematic error of the weighing process, such as the variation of the gravitational field at different places on the earth's surface, would have to be explicitly stated.)

The principle of accumulation is compatible with an individual coming to be constituted of the same quantity for a second time. Suppose, for example, the quantity in question is the wood of the timbers constituting the original ship of Theseus. As planks are changed, the constituting material during this time becomes ever greater. After yet more time, all the planks are exchanged, but the initial planks have been meticulously preserved. Finally, the new planks are all replaced by these original planks. For a time just after this replacement has been made, the constituting material is a quantity identical with the constituting quantity shortly after the ship was first built. Although artifacts may be susceptible to such possibilities, the corresponding situation doesn't seem to arise in the course of nature for naturally occurring objects, which are the primary focus here. But I don't think that the possibility is ruled out by any principle governing the workings of nature.

Although fixed for a fixed time, the material constituting an individual can vary from time to time. This allows that the material of two individuals might be successively interchanged during some time t, so that although each is constituted of different matter at shorter subintervals of t, they are constituted of the same matter at t. Examples of this kind are counterinstances to a criterion of identity in terms of sameness of constitutive matter, at least on one understanding of this condition. But the notion of constitution can serve as the basis of a closely related but more plausible criterion of identity requiring having the same matter throughout (at all subintervals of) any time, t. In other words, let the notion of the sameness of constitutive matter be understood by defining it as a three-place relation *SameConst* of having the same constitution between individuals x and y and an interval t when, for all subintervals t' of t, some quantity constitutes both x and y at t':

$$SameConst(x, y, t) \equiv \forall t' \subseteq t \exists \pi \, (Const(x, \pi, t') \wedge Const(y, \pi, t')).$$

Then the criterion of identity is: If x and y ever have the same constitution, they are identical, i.e.

(CI) $\quad SameConst(x, y, t) \supset x = y.$

By this criterion, there is no time throughout which distinct individuals have the same constitution. They are not constituted of the same matter for at least some part of any time, which is not to say that they are constituted of different matter for at least some part of any time—see next paragraph. The criterion allows that two individuals completely interchange their constitutive matter over a certain period and so are constituted of the same matter for that period if they don't have the same constitution for all subintervals of that period during which they are interchanging matter.

It might seem that the criterion is threatened with trivialisation in view of the position taken earlier that individuals, unlike quantities, are fleeting things which come into and go out of existence. Were it the case that individuals had the same constitution at times when they don't exist, then there would be just one individual. This is not a problem, however, since the definition of having the same constitution doesn't, as things stand, have this implication. But to be quite clear, a time-dependent existence predicate, $E(x, t)$, appropriate for transient beings is readily defined in terms of the constitution relation, as "x is constituted of some quantity at t". An individual's lifetime is then just the maximal period the individual is constituted of something, described by the relation L:

$$L(t, x) \;\equiv .\; E(x, t) \wedge \sim\exists t' \,(t \subset t' \wedge E(x, t')).$$

These definitions presuppose a *principle of restricted lifetime* which should be made explicit, that any individual is constituted of some matter for some time but not for any other times:

$$\exists t \,(\exists \pi \; Const(x, \pi, t) \wedge \forall t' \;(t \mid t' \supset \sim\exists \pi \, Const(x, \pi, t'))).$$

From these definitions and underlying principle it can be seen that at times when two individuals don't exist, they are not constituted of anything, and so not of a null quantity because mereology allows no such thing.

The identity criterion is correct, it seems to me, as it stands (in so far as individuals are considered material objects), but a possible strengthening of the condition (i.e. a weakening of the criterion—of the implication as a whole) might be considered. The existential quantification in the antecedent cannot be simply strengthened to universal quantification since it wouldn't then apply to any individual because individuals, as just explained, are not permanent existents. But the criterion can be weakened by requiring sameness of constitution for a common lifetime:

$$L(t, x) \wedge L(t, y) \wedge SameConst(x, y, t) \;.\!\supset\; x = y.$$

Distinct individuals would have different lifetimes or be constituted of different matter for at least some time during their lifetime. For the purposes of comparing this second, weaker criterion with the first, stronger criterion (CI), consider a scenario described by Simons (1987, pp. 201–2) of a variation on the ship of Theseus where another vessel is reconstructed from the original planks, which have been saved after successive replacements have resulted in complete refitting of the original ship with new timbers. Simons suggests that these two vessels coincide up to the first refit because they are made of the same material. But the first criterion wouldn't allow that two vessels are constituted of the same material throughout a time, and would therefore suggest that a new vessel is created after the refitting of the original ship. It would be natural to say that the refitted ship is identical with the original one, and the new ship is the one subsequently made from the original timbers. The second criterion doesn't rule out the possibility of there being two ships constituted of the same matter for a part of their lifetimes provided that this is a proper part, so that they are distinguished either by different lifetimes or, as in Simons' scenario, by being constituted of different matter at other times.

The identity criteria at issue here are not independent of how the constitution relation is applied (nor is the application of this relation independent of the principles governing it, including the identity criterion). When an individual begins to exist or whether it ceases to exist when its constitutive material is changed rather than continuing to exist are questions concerning how the constitution relation is applied. An ordinary ship may be deemed to continue to exist when its planks are exchanged for new ones, but not a ship of historical interest. Extensive efforts are made to preserve the material of the warship Vasa, which founded in Stockholm harbour in 1628 but was salvaged in 1961 and is now housed in the Vasa Museum. Had it collapsed during the salvage operation or been subjected to complete replacement of planks before being raised, and the original planks were then reassembled, constitution of the original material would no doubt be deemed to determine what is Vasa. Yet a third interpretation consistent with (CI) would be to say that the first ship ceases to exist when all its material has been replaced and a new ship sees the light of day alongside the other new ship constructed from the old material. The notion of an individual is a very broad one, and the identity criterion (CI) determines distinctions up to a certain point, definitely eliminating only proposals inconsistent with it and perhaps leaving several alternatives which can only be eliminated by more specific considerations of the kind of individual at issue.

In this example of Simons' I favour the ruling given by the stronger, first criterion, which is simpler, although the difference may not be very significant if it only turns on unusual situations realised by artifacts. An example not involving artifacts which is something of a time-reversed analogue of Simons' example would be two pools of water which grow as the rain continues and eventually merge, yielding a single large puddle. On the stronger criterion (CI), the analogous interpretation would be that one of the two pools consumes the other, which ceases to exist after the fusion. But we might equally say that both pools cease to exist and a new puddle is created by the fusion, or even that there is just one puddle, which is scattered at first but becomes connected after fusion. Perhaps the range of seemingly equally acceptable alternatives is indicative of the feeble claim a pool of water has to being rated an individual rather than merely an accumulating mass of material—neither a quantity nor yet a fully-fledged individual—for which no general term has been introduced here. But as with the previous example, the identity criterion allows but doesn't determine what is the most natural interpretation, which depends on other factors. Adopting the weaker identity criterion of having the same constitution throughout the same lifetime would allow the possibility of the two pools combining to form a single pool after merging, analogous to Simons' ship.

Something like the constitution criterion is the first of what Ayers thinks of as two equivalent identity criteria:

> The traditional view that individual[s] ... are distinguished at any one time by their matter can be expressed in two equivalent ways. First, as the principle that two distinct individual[s] ... cannot be composed of the same matter at the same time. Second, as the principle that, if two individual[s] ... have at any time all parts in common, they are identical. (Ayers 1991, p. 84)

But as noted above, the condition of being constituted of the same matter must hold for all subintervals of the time in question. Perhaps Ayers was thinking of times as instants, although I hardly think that the traditional view would be committed to this. The second principle is often denied, Ayers goes on to say, by modern philosophers who radically reinterpret it "by introducing the paradoxical notion that material objects have temporal parts". No such radical reinterpretation is advocated here. Although a notion of time-dependent parthood is introduced below, it is that of temporary parts (cf. Simons 1987, p. 175) and not the four-dimensionalist's notion of temporal parts that Ayers objects to. Nevertheless, I wouldn't accept the second formulation literally as an equivalent formulation of the first because constitution is not mereological parthood and no part of an individual's constituent matter is counted part of the individual. An equivalent formulation of the criterion of identity is to be had in terms of the notion of occupying a region of space, which is discussed in the next section.

3.3 Occupation for Individuals

The relation of occupying between a quantity of matter, a region and a time has been taken as primitive. Since an individual is constituted of a definite quantity at a given time, a natural thought might be that the region occupied by an individual at a time is that occupied by its constituent matter at that time. While this might be appropriate for individuals with unchanging constitution, it isn't for typical individuals which vary their constitutive matter. A river, for example, is ordinarily thought to occupy a region delimited by its banks, the river-bed and the water level. But in view of the principle of accumulation for the constitution relation, the definition just given would always make a river with any flow at all occupy a region exceeding this. Accordingly, an individual's occupying a place can't be reduced to what its constituent material occupies, and will have to be treated autonomously.

In the interesting cases, then, there is no question of an individual coinciding with its constitutive matter. But authors who find coincidence problematic will, no doubt, still object to spatial overlapping, for example a river's occupying a proper part of the region occupied by its constituent matter for a given time. Thus, Sider (2001, p. 141) conjectures, "The actual laws of nature presumably prohibit interpenetration of distinct material objects". Whether this is true of the thermodynamics of mixtures or the quantum mechanics of bosons is not so obvious, and Sider doesn't go into any more detail. But in any case, it is unclear what bearing it has on objects distinguished as individuals occupying part of the region occupied by their constitutive matter. Presumably, to use Sider's word, the laws he has in mind apply to the constitutive matter. The claim here is not that two individuals occupy overlapping regions, or that two quantities occupy overlapping regions, but that an individual and its constitutive matter occupy overlapping regions. The antagonist must engage with this claim, show that physics states laws about individuals and not just about quantities of matter, and show how these laws preclude the kind of spatial overlapping actually at issue here.

Rivers don't have to be taken seriously when doing pure chemistry, concerned as it is with an ontology of quantities and an ideology of substances, phases and conditions affecting the possession of such features. But for ecology and geography, and other disciplines for which they are significant, rivers do have to be taken seriously. No general thesis of the reduction of special sciences to physics is assumed here. In particular, it seems that the notion of a river's occupying a place can't be reduced to what its constituent material occupies, and must be treated autonomously. But even if the region occupied by an individual is not simply the region occupied by its constitutive matter, occupying is closely tied to being constituted of, and thus to time-dependent existence as the notion was introduced in the previous section. The general principle connecting the two concepts, which I refer to as the equivalence principle, is: When individuals x and y occupy the same place, they are constituted of the same matter, and conversely. Formally,

(EP) $\exists p\, (Occ(x, p, t) \wedge Occ(y, p, t)) \;\equiv\; \exists \pi\, (Const(x, \pi, t) \wedge Const(y, \pi, t))$.

The use of distinct variables doesn't, of course, imply that they refer to distinct objects, and where $x = y$ this reduces to: x occupies some place at t iff it is constituted by some quantity of matter at t. Occupying somewhere at t is therefore equivalent to existing at t as this was defined in the last section. It follows from this that the principle of restricted lifetime in the last section can be equivalently formulated as

$$\exists t\, (\exists p\, Occ(\pi, p, t) \wedge \forall t'\; (t \mid t' \supset \sim\exists p Occ(\pi, p, t'))).$$

This stands in contrast with what holds for quantities, which always occupy somewhere:

$$\exists p\, Occ(\pi, p, t).$$

The point of the more general formulation of the equivalence principle is that it naturally yields an equivalent formulation of the identity criterion for individuals presented in the last section, as I will now argue.

Rivers swell and contract with the seasons, and over longer periods even change course, either naturally or as a result of human intervention. Plants and animals grow, and move or are moved around. Think of a weed chased around the garden by the gardener pulling up shoots from one end of a root and stimulating it to grow at the other. So the time-dependency is just as much a feature of occupancy for individuals as for quantities. The times at issue are still intervals, so account must be taken of the possibility of movement during the interval, and the *principle of accumulation for occupation* introduced in Sect. 3.3 of the last chapter is repeated here:

(AO) For any time t, if an individual occupies a region p at t, then that region is the mereological sum of all regions occupied by that individual for some subinterval of t.

If a river's water were completely stagnant with no evaporation or supplementation by, for example, rainfall for a certain time, the region occupied by the river would be the same as that occupied by its constituent water during that time. But such circumstances never obtain, if only because there is a continual exchange of water between the liquid in the river and the water in the gas phase in the atmosphere. Accordingly, a river doesn't ever occupy the same place as its constitutive water. It might conceivably cooccupy with a proper part of the constituent water, namely a part which occupies a region at the source-end of the river at the beginning of the time in question and which just reaches the estuary at the end of the period. But this assumes no turbulence along the course of the river intermingling quantities of water as they transverse the river's length, and no evaporation or exchange with the gas phase making the quantity occupy a larger region than that occupied by the river, or rainfall making the quantity occupy a smaller region.

Therefore individuals and their constituent matter spatially overlap and in the extreme case, coincide (occupy the same region) at the same time. The same cannot be said of pairs of individuals, which are subject to a restriction on cooccupancy. The appropriate principle must take account of the time at issue being an interval with proper parts. Over sufficiently long times, the regions swept out by two individuals in motion may well overlap and even be identical—two test tubes counterbalancing one another as they rotate in a centrifuge, for example. But then if shorter intervals are considered, the regions swept out merely overlap, and for sufficiently short intervals, don't overlap at all. Individuals x and y will be said to *coincide* at t if at all subintervals of t they cooccupy some region, i.e. if they occupy the same region as each other at each subinterval of t. The order of the quantifiers is important in the definition of the *Coin* relation:

$$Coin(x, y, t) \equiv \forall t' \subseteq t \, \exists p \, (Occ(x, p, t') \wedge Occ(y, p, t')).$$

For a fixed time, the coincidence relation so defined is clearly transitive in view of the assumption of the uniqueness of the region occupied by an individual at a time, and it is clearly reflexive (for fixed t).

Remembering Simons' example discussed in the last section involving the building of a new ship from the planks removed from the original ship of Theseus, it was argued against the possibility raised by Simons that the two ships are made of the same material before the second building operation by appeal to the stronger criterion of identity. This was the criterion (CI), to the effect that individuals having the same constitution throughout some time are identical. It could equally be said that individuals that coincide throughout some time are identical, so that there are not two coinciding ships prior to the second building as Simons would have it, just one. In fact, in view of the equivalence principle, the notion of having the same constitution throughout some time is equivalent to that of coinciding throughout this same time, so that from the first criterion of identity it follows that individuals coinciding throughout some time are identical, and vice versa. The restriction on cooccupancy yields, therefore, an equivalent criterion of identity:

$$Coin(x, y, t) \supset x = y.$$

There is no time throughout which distinct individuals coincide: at any time, either they don't exist (don't occupy any region) at that time or they don't occupy the same region throughout that time.

Just as the first identity criterion could be weakened by restricting the times to the individuals lifetimes, so the equivalence principle yields an equivalent weakened criterion of identity in terms of coincidence throughout the individuals' lifetimes. But this is not adopted here.

3.4 Component Parts

There is another way of considering at least some sufficiently complex individual's make-up, in which physiology rather than chemical constitution is of primary interest. Individuals, or at least some of them, might be considered to have a structure of components, themselves individuals constituted of matter, arranged and joined in such a way as to form the individual in question. A component part of an organism might be an organ, for example, and a component part of the earth is the lithosphere. Like constituent matter, an individual's components might change from time to time, so that x could well be a component of y for only a part of y's lifetime. The relation of being a component part is therefore a time dependent one.

The question immediately arises of how the components are related to the constitutive matter. A summation principle suggests itself to the effect that an individual's constituent matter is the sum of the constituent matter of each of its components. But it is not obvious that the constitutive matter of the component parts jointly exhausts the matter of an individual in this way. The air under a car bonnet at t does not constitute the material of any component of the car. Is this air reasonably discounted as part of the material constituted by the car at t? What about the lubricating oil? These stuffs may not be necessary for a car on display; but they are for a car running normally. Similar considerations apply to the petrol in the tank, the oxygen needed for its combustion and the water and glycol antifreeze in the cooling system. It is difficult to see how they could be excluded from the constitutive matter of a car at a time t when it is speeding along the road, although they don't seem to form part of the constitutive matter of what are ordinarily called components. But perhaps construing components as solid, which is what leads to these problems, is a special case which shouldn't be taken as typical. The solidity apparently required of car components is not required of the organs of the human body, one of which is the lympho-reticular system, which includes the liquids blood and lymph (as well as the bone marrow, parts of the liver, and so forth). Suppose, for the sake of the following discussion, that components can be so construed to ensure the truth of the summation principle by encompassing all the constitutive matter of the individual in the constitutive matter of the components. If any criterion of identity formulated in terms of components is not to conflict with the criterion already formulated in terms of constituent matter, then something like the summation principle is crucial. Perhaps the moral to draw is that the criterion of identity in terms of constituent matter is fundamental, and doesn't in general yield one in terms of components. This should appeal to those who think that some sort of circularity afflicts identity criteria for a kind of entity formulated in terms of the same kind of entities.

The physiology of an individual introduces a level of structure over and above its chemical constitution. Thinking of this structure as spatial, it is natural to wonder whether it is possible to delimit the notion of a component in terms of the apparatus already introduced. In fact, a time-dependent analogue of the mereological part relation, naturally thought of as spatial parthood, can be defined in terms of the occupancy relation, for which corresponding analogues of some of the basic features of order relations exhibited by mereological parthood can be established. Spatial parthood between individuals might be considered a natural interpretation of the component relation introducing constituent parts.[2] At any rate, component parts are naturally thought of as spatial parts, and since spatial parts as defined here will be individuals, the converse might be more readily accepted. Component parts are the "other countable 'things'" that Prior says countable things are made from.

A first shot at defining "x is a proper spatial part of y at t" might run "the region occupied by x at t is a proper part of the region occupied by y at t", in which the time-dependency, lacking in the dyadic proper part relation of mereology standing between the regions, is introduced via the occupancy relation. But this would count the smaller of two bodies encircling one another as a proper spatial part of the larger one when it sweeps out a region at t which is a proper part of that swept out by the larger one. For sufficiently short times, however, the regions swept out don't overlap. A time-dependent relation of proper spatial part between individuals which wouldn't count the smaller of the two bodies in this example as a spatial part of the larger one can be defined by considering all subintervals of the time in question. Accordingly, "x is a proper spatial part of y at t", symbolised $x \sqsubset_t y$, is defined as "for all subintervals t' of t, the region occupied by x at t' is a proper part of the region occupied by y at t'":

$$x \sqsubset_t y \equiv \forall t' \subseteq t \, \exists p \, \exists q \, (Occ(x, p, t') \wedge Occ(y, q, t') \wedge p \subset q).$$

As already noted, an individual's occupying somewhere at a time implies existing at that time in view of the equivalence principle and the definition of time-dependent existence for individuals. The definition of individuals standing in the proper spatial part relation at a time therefore implies that they both exist at that time. And given the assumption of the uniqueness of the region occupied by an individual at a time, transitivity (for fixed t) of component (spatial) parthood follows from that of the mereological proper part relation. Irreflexivity follows similarly, and so proper component parthood is asymmetric.

[2]Spatial parthood provides an interpretation of what Simons (1987, p. 158) understands as constitution when he says "constitution is transitive". Holding times fixed, constitution as understood here is not transitive because it is always a quantity which constitutes an individual.

Pairing proper component parthood with the transitive and symmetric relation of coincidence defined in the last section, a more general reflexive and transitive relation of component parthood, $x \sqsubseteq_t y$, can be defined as "x is either a proper spatial part of or coincident with y at t":

$$x \sqsubseteq_t y \equiv . x \sqsubset_t y \vee Coin(x, y, t),$$

where $x \sqsubseteq_t y$ is read "x is a component (spatial) part y at t". This implies the existence of x and y at t since both disjuncts do. A symmetric relation, $\oplus_t$, of spatial overlapping can then be defined by analogy with mereological overlapping as the existence of a common component,

$$x \oplus_t y \equiv \exists z (z \sqsubseteq_t x \wedge z \sqsubseteq_t y),$$

and implies the existence of each of the overlapping individuals at the time in question.

Time-dependent analogues of the mereological relations can thus be defined with the appropriate properties. What about corresponding analogues of the mereological operations? This raises questions of existence which can't be settled simply by definition. But we might tentatively consider how a time-dependent analogue of mereological summation would be introduced.

3.5 Operations on Component Parts?

Following the way summation is defined in standard mereology in terms of overlapping, a time-dependent sum operation, $S_t x\, \varphi(x)$, of φ-ers at time t would be defined along the lines of

$$\exists! \varphi(x) \supset . y = S_t x\, \varphi(x) \equiv \forall z (z \oplus_t y \equiv \exists x (\varphi(x) \wedge z \oplus_t x)).$$

Because there are no null elements in mereology, an existential clause ensuring that there are φ-ers at time t would have to be satisfied. This is written here as $\exists! \varphi(x)$, leaving the question of its specific form aside for the moment. If the definition is to be justified, it would have to be established, given the existence of φ-ers at time t, both that there exists an entity y such that $\forall z (z \oplus_t y \equiv \exists x (\varphi(x) \wedge z \oplus_t x))$ and that this entity is unique. Following standard mereology, where the existence clause is laid down as an axiom schema, what would be required in the present case is something along the lines of

$$\exists! \varphi(x) \supset \exists y \forall z (z \oplus_t y \equiv \exists x (\varphi(x) \wedge z \oplus_t x)),$$

which would take a precise form once $\exists! \varphi(x)$ is clarified. Uniqueness in the case of standard mereology is established from the properties of the mereological relations deriving from the other axioms. In the present case, uniqueness would have to be established in one way or another. Finally, since what is at issue is an operation on individuals, time-dependent sums of individuals would themselves be individuals.

The open formula $\varphi(x)$ in the existence prerequisite, $\exists!\varphi(x)$, is an arbitrary formula, and might express some condition on x obtaining well before the time t. Suppose, for example, it is the condition of being a dinosaur, and t is the present. In view of the temporal existence implications of spatial parthood, no dinosaur is a spatial part of, or spatially overlaps, anything now. Accordingly, the existence prerequisite $\exists!\varphi(x)$ cannot simply be $\exists x\, \varphi(x)$, as it is in standard mereology, but should be complemented with a temporal existence condition along the lines of $\forall x\, (\varphi(x) \supset E(x,t))$. This is not quite subtle enough, however.

Just as an individual may well be constituted of different matter at different times, it may well survive a change of components. A forest survives the refurbishment of component trees by others as they die and decay and are replaced by new growth. Whereas quantities of matter don't go out of existence, components are individuals which may well do so. Over a sufficiently long period, an individual's initial components may have all been removed and destroyed by the end of its lifetime, having been exchanged for new components which were not in existence at the beginning of its lifetime. In order to accommodate this, the tentative definition of a sum must be modified and the existence prerequisite amplified. As it stands, the definition only sums over individuals which are overlapped throughout the time t to which the sum is relativised—it is whatever is overlapped throughout t by whatever overlaps some φ-er throughout t (cf. definitions of the spatial relations of part and overlap). This would exclude components coming into existence after the beginning of this period or going out of existence before the end of this period. A qualification must be introduced quantifying over subintervals of the time of summation along the following lines:

$$\exists!\varphi(x) \;\supset . \; y = S_t x\, \varphi(x) \equiv \exists t' \subseteq t\, \forall z\, (z \oplus_{t'} y \equiv \exists x\, (\varphi(x) \wedge z \oplus_{t'} x)). \qquad (3.1)$$

A corresponding modification of the additional condition mentioned in the last paragraph, $\forall x\, (\varphi(x) \supset E(x,t))$, to be incorporated in the existence requirement $\exists!\varphi(x)$ is also required. It should (i) be generalised to $\forall x\, (\varphi(x) \supset \exists t' \subseteq t\, E(x,t))$, allowing that the φ-ers don't necessarily exist throughout the time t to which the sum is relativised. Further, (ii) this should be supplemented with the condition $\forall t' \subseteq t\, \exists x\, (\varphi(x) \wedge E(x,t'))$ ensuring that there is always some φ-er during t.

The upshot is that time-dependent sums of individuals have something of the character introduced by the accumulation conditions on the matter constituting and on the region occupied by individuals during a period of time. The sum would include everything that would have been included in sums taken over subintervals of the time in question, even though these things may not exist throughout the time over which the sum is taken. Such individuals are not component (i.e. spatial) parts of the sum for the time to which the sum is relativised, however, because spatial parthood at a time entails existence at that time. In fact, such time-relativised entities as these sums are not individuals as these have been understood here.

Individuals are temporary existents, first presented as having varying constituent matter, then as having varying component parts. The varying constituent matter was taken as one reason for not identifying individuals with quantities of matter. Likewise, because of their varying components, an individual can't be identified with the sum, for some appropriate time, say its lifetime, of its component parts. Since sums may be different for different times, some definite time must be settled on for a putative identification with an individual. The lifetime of the individual seems the least arbitrary choice to make. But this would include all the past, present and future components of the individual, and wouldn't itself be something with varying component parts. As something which survives change in its components, an individual is not a time-relativised object x-at-t,[3] as sums defined in terms of time-dependent analogues of mereological relations necessarily are. Whatever they are, these putative sums are not operations generating individuals from individuals, and so not sums properly so called.

The difficulty that the sum includes things which are not spatial parts at the time in question would be circumvented by a different conception of sum in accordance with the original suggestion, without introducing the existential quantification over subintervals in (3.1) and without the corresponding complications in the existential prerequisite $\exists!\varphi(x)$. On this simpler conception, sums would include only those things x satisfying the defining condition $\varphi(x)$ throughout the time to which the sum is relativised. But they would still be time-relativised entities, which individuals aren't, and thus not the result of an operation generating individuals from individuals. They certainly wouldn't be things which change their component (spatial) parts from time to time.[4]

There doesn't seem to be a viable sum operation standing to the time-dependent analogues of the mereological relations as mereological summation stands to the mereological relations. Analogues of other operations ordinarily defined in terms of the sum operation in standard mereology fall by the wayside along with the time-dependent sums. In particular, the absence of a difference operation leads me to agree with van Inwagen (1981) in rejecting something like what he calls "The Doctrine of Arbitrary Undetached Parts" or DAUP for short, understanding in his statement of the doctrine,

[3]This would remain true even if individuals were defined, for the purposes of reducing the four-sorted language to a single-sorted language, in terms of constitution as entities x such that for some quantity π and some time t, x is constituted of π at t, because of the existential quantification.

[4]The thought that individuals are not sums of their components was captured by the seventeenth-century philosopher Stephanus Chauvinus: "We call properly *individual*, not what cannot in any way be divided, but what cannot be divided into several individuals of the same kind as itself or into several individuals specifically similar to him" (quoted by Chauvier 2017, p. 6). Complex biological organisms like dandelions, with a composite flower consisting of many tiny flowers or florets, and Portuguese men o'war, which are integrated colonies of specialised organisms called polyps, are not a sum of such components because they are not dandelions or Portuguese men o'war.

> for any proper part of a material object like Descartes, there exists an object which is the material object in question less the proper part.

that the notion of a proper part at issue would be that of a proper component or spatial part. Van Inwagen illustrates what he sees as the problem by supposing that Descartes looses his left leg and that there is a remainder, Descartes less his left leg, which he calls Descartes-minus. Both Descartes and Descartes-minus survive the amputation, from which it seems natural to conclude that Descartes and Descartes-minus are identical. But before the amputation it seems equally natural to say that Descartes and Descartes-minus are not identical. "At *t*, Descartes lost L [his left leg] and lost no other parts (save parts of him that overlapped L)" (van Inwagen 1981, p. 126), where *t* is the time of amputation. So before *t*, Descartes had a part which is not a part of Descartes-minus, which is why they are not identical. But this is in direct contradiction with what the principle of the transitivity of identity apparently requires:

Descartes = Descartes-minus + his left leg.
Descartes = Descartes-minus.
∴ Descartes-minus = Descartes-minus + his left leg

This latter argument is to be rejected because the identity statements are improper: there is no sum or difference operation generating objects which could stand in an identity relation to anything. But this is probably not how van Inwagen would put it.

Van Inwagen feels he can't solve the problem by denying the existence of Descartes-minus without drawing the "even more striking conclusion" that Descartes's left leg doesn't exist either:

> there was never any such thing as Descartes's left leg. We need only one premise to reach this conclusion, namely that if L existed, D-minus did too. And this premise seems quite reasonable, for it would seem wholly arbitrary to accept the existence of L and to deny the existence of D-minus. In more senses than one, L and D-minus stand or fall together. If these things existed, they would be things of the same sort. Each would be an arbitrary undetached part of a certain man. This fact may be disguised by our having (problems of multiplicity and vagueness aside) what is a customary and idiomatic name for L if it is a name for anything: "Descartes's left leg." But this is a linguistic accident that reflects our interests. (We may imagine a race of rational beings who raise human beings as meat animals. Suppose these beings, for religious reason, never eat left legs. They might very well have in their language some customary and idiomatic phrase that stands to D-minus in the same relation as that in which the English phrase 'Descartes's left leg' stands to L.) (van Inwagen 1981, p. 196)

But we might well take a different line on this. Time-dependent analogues of the mereological relations can be accepted without taking the further step of accepting corresponding analogues of the mereological operations. We can see, then, that Descartes's left leg might well be thought to exist during Descartes's lifetime as one of his components, but this need not imply that the remainder of Descartes's body amounts to a component and the existence of what van Inwagen calls Descartes-minus. This is to be distinguished from the matter comprising the remainder of Descartes's body once the matter constituting his left leg for some chosen time has been subtracted (bearing in mind that the latter will include matter such as the blood circulating thought both the left leg and the rest of the body during this time), over which quantities a generally viable difference operation is defined. It is therefore not "wholly arbitrary", as van Inwagen claims, to accept the existence of the leg but not the difference, Descartes-minus.

Some of the views van Inwagen has since developed on relativising mereological concepts to a time, which are even further at variance with those put forward here, are discussed in the appendix to this chapter.

The question of the summation principle discussed at the beginning of this section has not been completely resolved and the criterion of identity in terms of constitutive matter stands. It remains to be seen whether this is equivalent, for individuals with component parts, to a criterion in terms of sameness of component parts.

One reason for seeking a sum operation over individuals might derive from a wish to recognise non-connected individuals like the solar system, whose scattered components we might like to think of as bound together by an operation generating an individual from individuals. Such individuals provide one kind of counterexample to the strategy of reducing individuals to connected quantities of some sort. But in the absence of a suitable sum operation, we should just recognise that some individuals are like the solar system, with scattered components. In this connection, lack of an analogue of mereological summation for individuals means that there is no analogue of the principle that the region occupied by a sum of quantities is the sum of the regions occupied by those quantities (Sect. 2.4). Some such idea might nevertheless be feasible in terms of a suitable notion of a collection of individuals. Plural logic has been developed to accommodate a notion of a plurality, distinct from that of a set or a mereological sum, which is at issue in saying, for example, that the children encircled the bonfire, the boxes take up most of the space in the attic or the planets and asteroids form the solar system. The claim that a plurality, X, occupies a place p at a time t, $Occ(X, p, t)$, which is the sum of the places occupied by each of the individuals, x, that is an X (written $x \varepsilon X$) could then be expressed as

$$Occ(X, t) = \Sigma p \, \exists x \, (x \varepsilon X \wedge Occ(x, p, t)).$$

Again we see how convenient it would be if general summation over regions were allowed, in the absence of which we have to envisage an iteration of binary sums over the regions occupied by the individuals of the plurality taken in some order. Lack of an analogue of mereological sum for individuals doesn't, therefore, leave us without a notion of a collection of individuals. The plurality notion would seem to be the appropriate one. But this relies, of course, on not identifying pluralities with sums.

By way of a final word on this discussion, individuals are understood to be continuants, bearing properties and enduring for some period, to be constituted of matter, occupy spatial regions, enter into numerous other relations and participate in processes. But no simple, general criterion of what systematically distinguishes individuals from non-individuals, such as spatio-temporal boundedness or independence, has been offered. We might hope that in a restricted domain such as biology the notion of an organism could be delimited as a special case of an individual in the light of the theory of evolution as the bearer of traits that have been selected for the advantage they bestow on the individual, and not to any component part or more inclusive aggregate. But even here an examination of recalcitrant cases in the face of a multiplicity of well-motivated but mutually inconsistent proposals shows that

> We know that counting particular lumps of living matter, and not others, allows us to describe and make predictions about evolutionary processes. Yet we lack a theory telling us which lumps to count. (Clarke 2010, p. 323)

The present notion of an individual is considerably broader than that of a biological organism. But it seems to be presupposed by the notion of an organism. And just as failure to precisely delimit the concept has not prevented philosophers of biology from arguing for the centrality of the notion of an organism in biology, I don't see that problems with adjudicating between borderline cases detracts from the centrality of the notion of an individual in the ontology of continuants.

3.6 Modality

The proponents of the "constitution is identity" thesis constrain their discussion of examples to suit their view, choosing objects with unchanging constitution and leaving only modal properties to distinguish them. Katherine Hawley, we saw, goes so far as to consider the fantastic case of a statue and its constituent clay being created and subsequently destroyed at the same times. Whatever possible world this situation might obtain in, it is difficult to see what bearing it could have on the character of ordinary objects. More measured modal claims concerning what is possible under feasible circumstances are more interesting.

The fact that individuals typically undergo a change of constitutive matter with time is naturally understood to have a modal correlate. Individuals might well have been constituted of different matter from that of which they are actually constituted. I leave it open whether there might be individuals with a necessary constitution (perhaps historically valuable artifacts like the warship Vasa), but an organism like a plant, for example, would be constituted of different matter from that now constituting it if it had been moved and consequently carried out its photosynthesis processes in a different region of the atmosphere. A given individual might have had a longer or a shorter lifetime than it had. The possibility of a longer lifetime would entail not merely that it might have been constituted of a different quantity from that which in fact constituted it, but that it might have been constituted of something when it wasn't in fact constituted of anything. A possibly shorter lifetime would imply that it might not have existed when it actually did. It would seem that entire individuals might not have existed at all, or that there might possibly be individuals that don't ever actually exist, such as Wittgenstein's possible son or perhaps a machine that will transform the power of waves into a useful energy source.

Quantities, by contrast, are permanent entities that do not gain or lose parts. The modal correlate of this is that they have their parts essentially, with no possibility of gain or loss:

MEss (i) $\pi \subseteq \rho \supset \Box(\pi \subseteq \rho)$ and (ii) $\pi \nsubseteq \rho \supset \Box(\pi \nsubseteq \rho)$.

In fact, quantities are entities which can, in accordance with Lavoisier's principle of the indestructibility of matter, be neither created nor destroyed. This is a modal thesis, stating not merely that no quantities of matter ever come into existence or are destroyed, but that it is not possible that there are quantities which don't actually exist, or that quantities don't exist which actually do. Perhaps it could be stipulated that quantities are destroyed or come into existence in fairy-tale worlds as some sort of metaphysical possibility. But such scenarios hardly tell us what is possible for matter and fall outside the scope of the modality envisaged here. Quantities comply with what is called the single domain interpretation of quantified modal logic, and satisfy the Barcan formula,

BF $\forall\pi\,\Box\varphi \supset \Box\forall\pi\,\varphi$, equivalently, $\Diamond\exists\pi\,\varphi \supset \exists\pi\,\Diamond\,\varphi$,

and its converse,

CBF $\Box\forall\pi\,\varphi \supset \forall\pi\,\Box\varphi$, equivalently, $\exists\pi\,\Diamond\,\varphi \supset \Diamond\exists\pi\,\varphi$.

Times and regions of space are, like quantities, permanent objects which are never created or destroyed. The Barcan formula and its converse therefore holds for entities of these kinds too, along with the principles of mereological essentialism.

Kripke devised his variable domain interpretation to invalidate the Barcan formula and its converse, neither of which he thought should be part of the basic structure of modal predicate logic. The present line of argument suggests that these considerations apply to individuals, to which the Barcan formula and its converse don't apply. Taking φ to be $\exists t\,E(x,t)$, where $E(x,t)$ is the time-dependent existence predicate defined in Sect. 3.2, the antecedent $\Diamond\exists x\,\exists t\,E(x,t)$ of the equivalent form of the Barcan formula could well be true—the actual existence of individuals shows that their existence is certainly possible. But we might want to allow that the consequent, $\exists x\,\Diamond\,\exists t\,E(x,t)$, is false—the universe might have developed in such a way that individuals were never formed from matter. Less drastically, without going to such extremes, the above counterexample of the machine that transforms wave power into usable energy might exist, but there is no individual that might be turned to such a purpose. The converse Barcan formula is even more obviously invalid when applied to individuals. A typical individual might well not exist: $\exists x\,\Diamond \sim \exists t\,E(x,t)$. But it couldn't possibly be the case that there is something that doesn't exist: $\Diamond\exists x\!\sim\!\exists t\,E(x,t)$.

Kripke's own suggestion of how to formulate the axioms corresponding to the variable domain interpretation is generally thought inappropriate, and the usual way of doing this is to adapt the principles of free logic in which the truth-functional tautologies, the reflexivity and substitution axioms for identity and the rules of modus ponens and generalisation are supplemented with the axioms and axiom schemas:

Q1 $\forall x \varphi \supset . Ey \supset (y/x)\varphi$
Q2 $\varphi \equiv \forall x \varphi$, provided x is not free in φ
Q3 $\forall x (\varphi \supset \psi) \supset . \forall x \varphi \supset \forall x \psi$
Q4 $\forall x Ex$
Q5 $\Diamond Ex$

The instantiation axiom of standard predicate logic is qualified with an existence predicate, Ex (which could be defined as $\exists t \exists \pi \, Const(x, \pi, t)$), in axiom Q1, in view of which the right-to-left implication of Q2 is no longer provable as in standard predicate logic and is laid down as an axiom for the purposes of modal quantification theory (ensuring that $\forall x \varphi \supset \exists x \varphi$ is a theorem), and an additional axiom Q4 governing the existence predicate is added. Q3 is as in standard predicate logic.[5] Axiom Q4 is not expressed by an open formula, Ex, because the free variables are held to range over a so-called outer domain of all possible objects which, in view of Q5, comprises possible (possibly existing) individuals. All entities, possible and actual, referred to by the free variables are referred to rigidly, i.e. satisfy the ridigity principles

R1 $x = y \supset \Box x = y$
R2 $x \neq y \supset \Box x \neq y$,

ensuring that a term's denotation is preserved necessarily and cannot be otherwise. As Quine pointed out, R1 is straightforwardly provable from the identity axioms and the basic rule of necessitation in normal modal logic. R2 is provable as a theorem in stronger modal logics which include the "Brouwer" axiom, and otherwise is laid down as an axiom. The rigidity principles imply that possibly distinct individuals are necessarily distinct by the contraposition of R1, $\Diamond x \neq y \supset x \neq y$, and R2. Similarly, $\Diamond x = y \supset x = y$ by the contraposition of R2 whence $\Diamond x = y \supset \Box x = y$ by R1. Corresponding rigidity principles for quantity, time and space variables follow in normal modal logic with the principles of standard predicate logic from the principles of mereological essentialism.

The application of these axioms can be illustrated by a puzzle originally raised by Wiggins (1976) as a way of questioning the usual treatment of necessity as a sentential operator. Formulating what he takes to be the true claim, "Cicero is necessarily a man", in terms of a sentential operator as $\Box M(a)$ and "Cicero is a man entails men exist" as $\Box (M(a) \supset \exists x M(x))$, it follows in normal modal logic that $\Box \exists x M(x)$. But it is patently false that men necessarily exist. Not wishing to reject the principle of existential generalisation, Wiggins proposes to circumvent the problem with an alternative treatment of necessarily as a predicate modifier. But we might instead question the usual formulation of the principle of existential generalisation when applied to individuals in modal contexts. The principle of existential generalisation in free logic is given by the contraposition of Q1, namely

[5] For completeness of the basic system of quantified normal modal logic (Garson 1984), a "mixing" rule must be added in addition to the sentential modal principles of system K, namely

If $\vdash \chi \vee \Box(\psi \supset . Ey \supset (y/x)\varphi)$ then $\vdash \chi \vee \Box(\psi \supset \forall x\varphi)$, provided x is not free in ψ.

$Ey \wedge (y/x)\varphi \;.\!\supset \exists x\, \varphi$. As this view would have it, if the second premise of Wiggins's argument is to be true, it should be formulated as $\Box\, (M(a) \wedge E(a) \;.\!\supset \exists x M(x))$, in which case the deduction no longer holds. An additional premise, $\Box\, E(a)$, would be required for the conclusion to follow, which is precisely the point at issue, that it is not necessary that Cicero exists.

How does Wiggins's argument fare when applied to quantities? An example might be

$$\begin{array}{ll} & \Box\, \mathit{Water}(\pi), \\ & \Box\, (\mathit{Water}(\pi) \supset \exists\pi\, \mathit{Water}(\pi)), \\ \therefore & \Box\, \exists\pi\, \mathit{Water}(\pi). \end{array}$$

The former objection to the second premise fails because with the Barcan formula and its converse holding for universal quantification over quantities, which cannot disappear or be created, there is no call to modify the principle of existential generalisation in standard predicate logic. But by the same token, the first premise is false. Unlike a man, who is necessarily a man, a quantity is not necessarily water in view of water's susceptibility to conversion into other substances in the course of chemical reactions. A quantity which is water doesn't disappear when it ceases to be water but becomes, in accordance with Lavoisier's principle, the bearer of different substance properties. The "Water" predicate is time dependent, applying to a quantity at a time. Similarly, what is not water might in fact be water as the result of chemical reactions, so that, whereas whoever is a man is necessarily a man, and whoever not not, neither of the following two claims hold:

$$\mathit{Water}(\pi,t) \supset \Box\, \mathit{Water}(\pi,t), \qquad {\sim}\mathit{Water}(\pi,t) \supset \Box{\sim}\, \mathit{Water}(\pi,t).$$

Still, the argument would seem to be valid. A similar application of the argument to quantities where the premises are true is

$$\begin{array}{ll} & \Box\, \mathit{Ponderable}(\pi), \\ & \Box\, (\mathit{Ponderable}(\pi) \supset \exists\pi\, \mathit{Ponderable}(\pi)), \\ \therefore & \Box\, \exists\pi\, \mathit{Ponderable}(\pi). \end{array}$$

Here the property of being ponderable (having mass) would seem to be an essential property, the second premise is true (existential generalisation and necessitation) and the conclusion is acceptable. In fact, $\exists\pi\, \Box\, \varphi(\pi) \supset \Box\, \exists\pi\, \varphi(\pi)$ is a theorem. The existence of any particular species, or organisms in general, may not be naturally necessary, but the very existence of matter is another thing.

Kripke's suggestion of the necessity of origin might be a general essential property of individuals, necessarily true of every individual that has it, as being ponderable is necessarily true of every quantity. A way of putting this within the present apparatus is to first define an *initial composition* of an individual by

$$\begin{aligned} \mathit{Init}(\pi,x) \;\equiv\; & \exists t\, \exists\rho\, (L(x,t) \wedge \mathit{Comp}(x,\rho,t) \wedge \pi \subset \rho \wedge \exists t'\, (t' \subset t\, \wedge \\ & \exists t''\, (E_A t'' t \wedge E_A t'' t' \wedge \mathit{Comp}(x,\pi,t')))), \end{aligned}$$

where $E_A t' t$ is the relation of t' being earlier than and abutting t. A quantity π would then be distinguished as the *original composition* of x, $\mathit{Orig}(\pi,x)$, if it is the smallest initial composition of x—that which is part of every initial composition of x. The assumption would be that each individual is formed during some initial interval of time from a minimal quantity of matter, any proper part of which would not be sufficient to compose the individual. The modal claim would then be that if the individual has an original composition, implying that it exists at the time when thus composed, then this quantity is necessarily the original composition of the individual. A plant might well be constituted of different matter than it is during its lifetime if it had been moved, but its original matter couldn't have been different. And an individual might have had a longer or a shorter lifetime than it in fact had, but this affects the possible time of its demise and not the time of its initial formation, which it necessarily fixed by its necessary composition of its original matter. Something more might be said about the kind or kinds of substance instanced by the original quantity and its parts in the specific case of each individual or perhaps kind of individual, but such details would go beyond the general principle for individuals of whatever kind. The case for such an assumption has often been reiterated for organisms and artefacts. Planets have recently been deemed to have acquired a minimal seize, requiring a minimal quantity of matter (sufficient to ensure a spherical shape by self-gravitation and gravitationally dominate their vicinity), in order to qualify as such. But whether the principle holds for everything that counts as an individual with a possibly varying composition is another matter, and not one I feel able to confidently assert.

Like Cicero, individuals don't in general exist necessarily, and it is not easy to think of any particular kind of individual that does. Not only might actually existing individuals not exist; it seems that we can entertain the idea of a merely possible individual, namely an individual x such that

$$\exists t \diamond E(x, t) \wedge \sim\exists t\, E(x, t).$$

The first conjunct is $\exists t \diamond \exists\pi\, Const(x, \pi, t)$ by the definition of the time-dependent existence predicate, and by CBF the latter implies (and together with BF is equivalent with) $\exists t\, \exists\pi \diamond Const(x, \pi, t)$. So there is (without qualification) a quantity of matter which a merely possible individual would be constituted of, and since there is only so much matter to go round, this imposes a restriction on the nature and number of individuals that would be jointly merely possible.

Many properties of individuals are either in fact properties of their constituent matter or depend on their constituent matter in some way. If I were to eat less, I would be constituted of matter with less mass, and thus be lighter than I am or otherwise would be. Unlike a quantity, then, an individual needn't be as heavy as it is, although this is not to deny that everything is necessarily as heavy as it is in the sense that it would be contradictory otherwise. That an individual might have had a longer lifetime than it had can be expressed along the lines of

$$L(t, x) \wedge \diamond L(t', x) \wedge t \subset t',$$

which, even if it doesn't capture the lengths of the intervals in question, does capture a prerequisite for time t' being longer than that of t in terms of a relation that holds rigidly between the times in question in virtue of the principle of mereological essentialism. Similarly, an individual's being possibly heavier than it is comes down to a rigid relation of proper parthood between quantities:

$$Const(x, \pi, t) \wedge \Diamond\, Const(x, \rho, t) \wedge \pi \subset \rho.$$

It will be interesting in the sequel to consider whether there are any such modal comparisons which do not come down to rigid mereological relations between the appropriate entities.

An individual that might at some time have different constitutive matter from that it in fact has isn't constituted of its matter necessarily. The identity criterion CI of being constituted of the same matter should certainly not be weakened to $\Box\, SameConst(x, y, t) \supset x = y$ otherwise little or nothing would satisfy it. Correspondingly, strengthening the identity criterion to $\Diamond\, SameConst(x, y, t) \supset x = y$ would be too easily satisfied because whether actual or possible, individuals are constituted from the same pool of matter and it can't be ruled out that, had the past of one individual been less fortunate, it might have provided the matter of another individual. This suggests that an additional principle such as that of common origin is needed to determine cases of possible identity.

3.7 Appendix: Van Inwagen's Understanding of Parthood as Time-Dependent

Mereological relations of parthood, overlapping and separation are dyadic. This leaves no possibility for changing of parts over time. Having seen what sort of entities mereological concepts are applied to here, this is just as it should be. Times obviously do not change their parts over time, and nor do regions of space. Quantities of matter conforming to the principles of elementary science, and in particular Lavoisier's principle of the conservation of mass, interpreted with Poincaré to mean that "Lavoisier [...] has demonstrated the indestructibility of matter by demonstrating the invariability of mass" (Poincaré 1913, p. 65) do not gain or lose parts with time either. But van Inwagen maintains the contrary. We might well suspect that this turns once again on an equivocation between a proper treatment of entities standing in mereological relations and material objects with changing constitutional matter. Perhaps the issue is purely verbal, with van Inwagen essentially referring to what is here called the constitution relation as mereological parthood. But taking the challenge at face value, van Inwagen defends his thesis by mounting an attack and claiming to show that the idea that mereological sums cannot change their parts is "groundless" (van Inwagen 2006, p. 614). This claim is distinctly less startling when we learn that he changes the basic mereological relations to 3-place relations with the extra place reserved for a time and introduces what he calls a sum-relation, which is related to a time and departs in essential respects from the mereological sum operation, notably by not being unique and so not an operation. Moving the goal posts reduces the likelihood of persuading the opposition unless there are convincing arguments for having to make these changes from reasonably neutral ground. But we will see that this so-called argument is shot through with his own presuppositions, and powerless against anyone not already committed to the same view.

Van Inwagen claims to show that every object is a mereological sum, so that if any object can change its parts, so can mereological sums. Then he disputes an argument claiming to show that no object can change its parts and concludes that speaking of mereological sums changing their parts is not to misapply the concept of mereological sum. But of course it is, since mereological sums are not one thing at one time and another at another. Van Inwagen is not really talking about mereological sums at all.

The first move is to claim that "everything has parts: itself if no others" (van Inwagen 2006, p. 620), from which it would uncontroversially follow that everything is a sum of its parts. The reflexivity theorem, $\forall x\,(x \subseteq x)$, is supposed to show that everything has parts. But as noted in Sect. 1.2, such a universally quantified formula only says that everything in the universe of discourse of the formalised theory has the feature in question. A perfectly coherent formulation of the idea that some but not all things obey mereological principles was presented in the first chapter. An argument would be required to show this was mistaken. But what van Inwagen offers is simply the bare claim $\forall x\,(x \subseteq x)$, that is, the assumption on his part that everything has parts. There is no trace of an argument here.[6]

The next stage of the procedure is an appeal to authority in the form of a quotation:

> It really is the most obvious common sense that a physical object can acquire and lose parts. Parthood surely is a three-place relation, among a pair of objects and a time. (Thomson 1983, p. 213)

But Thomson's example of a house loosing a brick or two is clearly that of an object with varying material constitution, and not of a quantity of matter. Van Ingwagen doesn't recognise this distinction, but proposes instead to come to grips with the circumstance that a continuant loses and gains what he wants to call parts by relinquishing the uniqueness characteristic of mereological sums and settling for a 3-place relation, "x is at t the sum of the ys" (within a framework of plural quantification). This allows that x at t is the same sum as x at t' even if the things summed in the sum x at t' differ from the things summed in the sum x at t. A brick house, for example, is said to be a sum of a given collection of bricks on Tuesday, but later, having lost a brick, this same house is the sum of the Tuesday bricks less

[6]We might wonder whether spaces and times are included in van Inwagen's universe of discourse and whether he takes his argument to show that mereological parthood between regions and between times is time dependent.

the lost one. What does "same sum" mean here? Van Inwagen defines "x is the same mereological sum as y" by

> x is the same mereological sum as y iff x is a mereological sum and y is a mereological sum and $x = y$. (2006, p. 626)

So the identity of van Inwagen's sums is not determined by a mereological criterion at all, but whatever determines $x = y$, the identity of continuants. What stronger indication could there be that it is not mereology that is at issue here?

The point becomes even clearer in the treatment van Inwagen offers as he proceedes to consider an argument purporting to show that objects cannot change their parts. The argument envisages an object α which is the "mereological sum" of A, B and C (which he writes by putting $\alpha = A + B + C$). After some time, C is annihilated, and the question posed whether α still exists. A + B exists, as it did before C's annihilation. But α wasn't identical with that then and cannot become identical with something else later. Van Inwagen identifies two premises in this argument which he says are doubtful, but considers only one, to the effect that before C's annihilation, A and B had a mereological sum. (We have seen that van Inwagen questioned this in his 1981 article. But he doesn't refer to that discussion in the 2006 article.) This, he supposes, is accepted on the strength of the principle (p. 629) that

> For any xs, if those xs exist at t, those xs have at t at least one mereological sum.

But this he rejects because he rejects arbitrary sums, and says (p. 630) that he therefore sees no reason to suppose that an object cannot change its parts. Since he claims to have shown that everything is a mereological sum, he further concludes that mereological sums can change their parts.

This argument that van Inwagen rejects is not one that would be advanced here in support of the claim that objects cannot change their parts because no entities envisaged here as forming sums come into or go out of existence with time, and in particular, none are subject to annihilation. But van Inwagen is sufficiently impressed that he takes refuting it to support his own thesis that objects can change their parts. Like others who put forward arguments of the same general kind, however, he is surprisingly evasive in his handling of time given his commitment to the time-dependence of mereological concepts.

Van Inwagen's presentation of the argument hangs on the role of a changed state of affairs before and after a certain time. Why, then, symbolise "the mereological sum of x and y" as "$x + y$", using the dyadic operator "+" if it is to be sensitive to the time when a sum is formed? Why not write something like $x +_t y$ instead? The suspicion arises that presentations of this kind of argument trade on ambiguities in the imposition of concepts of what van Inwagen calls mereology. It would seem that anything identical with the time-relativised sum $A +_t B +_t C$ is a time-relativised object, α_t. And at t', after t, when C has, according to the tale, been annihilated, there is an object $\alpha_{t'} = A +_{t'} B$, time-relativised to the new time t'. It seems that there just isn't any object, α, plain and simple, of which we can ask whether it is both $A +_t B +_t C$ and $A +_{t'} B$ and complete the reductio argument. But van Inwagen insists that there is one and the same object, α, plain and simple (time independent, like a continuant), which is a sum $A +_t B +_t C$ and a sum $A +_{t'} B$, where $t' \neq t$.

I agree with van Ingwagen that standard mereology is not adequate to yield an account of changes undergone by individuals, but I see no reason why mereology should be massacred in the attempt to redress the issue. Perhaps he is essentially (small points of divergence aside) talking about what is here called the constitution relation. Mereology should be kept plain and simple, to be applied where appropriate. These gains should not be cast to the wind and a two-tier system of mereological concepts as originally conceived and modified mereological concepts should definitely be avoided. Issues which outstrip mereology should be met by additions to accommodate them.

Time clearly has a role to play in a reasonably full account of material objects—quantities and individuals. The question is how it is to be brought into the picture. Rather than modifying mereological concepts, which have such clear applications as originally conceived, alternative analyses are suggested here. Usually this will take the form of treating some non-mereological predicate as having place-holders for times. Suggestions along these lines have already been broached by speaking of constitution as a relation between an individual, a quantity and a time, which allows for the expression of the idea that individuals are constituted of different matter at different times. Material objects—quantities and individuals—might occupy different regions at different times. Chemical reactions mean that quantities might bear different substance properties at different times. We have seen here that van Inwagen proffers a different view, but has no independent arguments not assuming from the outset premisses that his antagonist would regard as contentious. It remains to see whether the opposing views offered here give a more plausible and complete picture.

Chapter 4
Distributivity and Cumulativity

4.1 Mass Predicates

Quantities of matter are distinguished from individuals by their mereological structure. This is reflected in the general features of certain predicates expressing properties of matter such as being of a kind of substance (e.g. "water") and exhibiting one or more phases (e.g. "liquid", "smoke"), which will be of particular interest in the following chapters. Such predicates are often called mass predicates in the philosophical literature after Otto Jespersen (1924) coined the term "mass word". The term alludes to the non-particulate, formless character of matter as it appears on the macroscopic scale, and writers on metaphysics often use the term in this sense, as when Holden (2004, p. 80) speaks of the structure of bodies as "a metaphysically simple (noncomposite) structure best described with mass terms" (taken up in Sect. 5.7 below). It would seem that it is to mass predicates we should look for general features of the quantities of matter constituting countable individuals.

Mass predicates are contrasted with predicates expressed by count nouns, which are said to individuate what they apply to. Count nouns occur in natural language in both singular and plural forms, the latter applying to a countable group of entities. "Apples", for example, applies to a countable number of fruits, and stands in contrast to predicates like "salt" and "brine", which don't have a plural form (with the same mass term sense, as opposed to, e.g., speaking of the halogen salts of the alkali metals) nor do they individuate what they apply to into one or more countable entities. We can ask How much? of what mass predicates apply to, but not How many?, associating measurability with mass predicates as countability is associated with count nouns. Humans comprise a lot of water, very little selenium and lots of blood but many blood vessels, a few limbs and only one heart.

However, a terminological confusion threatens. There are non-count nouns like "furniture", "clergy" and "information" that also fulfil grammatical criteria of the kind just indicated but which are clearly not mass predicates in the sense of

P. Needham, *Macroscopic Metaphysics*, Synthese Library 390,
https://doi.org/10.1007/978-3-319-70999-4_4

attributing a non-particulate, formless character to matter. Nevertheless, they are classified as mass predicates by linguists and linguistically inspired philosophers. The class of non-count predicates is clearly broader, but its actual extent is notoriously peculiar to particular languages. Whereas the English worry about "lots of false information", the French speak of "les fausses informations qui circulent". The German "Möbel" takes an indefinite article, whereas the Swedish "möbel" resembles the English "apple" in taking an indefinite article and the correlates of "many" and "few" when used in the plural rather than "much" and "little", although it can also be used (like "apple" and unlike the German "Möbel") in the singular with the correlates of "much" and "little". We can't ask How much? of clergy, and although it is grammatically correct to ask How much? of furniture, there is no real basis for measuring how much furniture except by counting the items or measuring the amount of their constitutive matter.

It comes as no surprise that linguists generally concur in rejecting a mereological analysis of mass predicates on their understanding of the term. Whether we should attach any metaphysical significance to this is another matter. We don't, in particular, have to agree with Gillon that a mereological analysis is inappropriate

> for reasons set out by Parsons (1970, section VI.A). Under the standard and plausible mereological assumption that two wholes are identical if and only if they have the same proper parts, it turns out that, in a world in which all furniture is made of wood and all wood has been made into furniture, the whole of wood would be identical with the whole of furniture—an implausible consequence. (Gillon 1992, p. 599)

Expressing his indebtedness to Quine, Parsons himself says "parts of chairs might be wood without being furniture" (1970, p. 377). But as argued here, the "made of" relation in which the wood is said to stand to furniture is not the mereological part relation but the "constitutes" relation discussed in Chap. 3. I have no qualms about the mereological criterion of identity, but items of furniture are to be distinguished from their constitutive wood, not to mention the changing moisture content, which varies as the wood dries out or absorbs water according to the ambient circumstances of the furniture in any possible world sufficiently like the actual world to be of interest.

It would be less confusing if superficial grammatical criteria were said to delimit a class of non-count predicates. But "mass predicate" and cognate terms are sometimes used in this broader sense and will for dialectical purposes be difficult to avoid in the following sections. At the end of the day, however, it is predicates describing matter as a substance, a phase or a mixture that are of interest here. Although I try on the whole to avoid it, where the dialectical context doesn't dictate otherwise the term "mass predicate" is used in the general metaphysical sense first outlined above of referring to the non-particular, formless character of matter rather than to refer to all and sundry non-count predicates.

Two general features of mass predicates (in the metaphysical sense of the term) arising from the mereological structure of what they apply to are of interest. A distributive and a cumulative condition have been suggested as jointly providing an analysis of the non-particulate, formless character ascribed by mass predicates.

Although upholding the cumulative condition, however, Quine has argued that the distributive condition is incompatible with modern views of matter and hinted at a restriction. But this restriction is inadequate and the feasibility of dropping the distributive condition whilst retaining only the cumulative condition can be questioned in the light of an adequate generalisation to accommodate polyadic predicates. These issues are discussed in this chapter together with the formulation of a restriction on both conditions to spatial parts. The applicability to these and related conditions to matter as understood on older and more modern views of matter is taken up in the following chapters.

4.2 The Distributive and Cumulative Conditions

Substance predicates, describing matter as of a particular kind of substance, would seem to share the general feature of applying to all of what they apply to. The idea goes back to Aristotle, who says, "any part of water is water" (*De Generatione et Corruptione* I.10, 328a10f.)—a quantity which is water is all water. Formally, a predicate $\varphi(\pi)$ is distributive if

$$\varphi(\pi) \wedge \rho \subseteq \pi . \supset \varphi(\rho).$$

A phase predicate like "liquid", apparently describing a condition of matter without individuating what it applies to, might also seem to comply with this condition. The word does take the indefinite article and have a plural form, but when speaking of a liquid or two liquids we are speaking of one or two kinds. Considered as a phase predicate, "is liquid" would seem, on the face of it, to be just as much a mass predicate as substance predicates like "is water", so that a quantity which is liquid is all liquid. As has already been suggested in Chap. 2 and will be taken up later, things are not what they seem. But we assume for the present that they do apply to fixed quantities of matter for some interval of time and conform to the distributive condition.

There is an objection to the distributive condition as applied to substance predicates. As Quine puts it, "In general, a mass term in predicative position may be viewed as a general term which is true of each portion of the stuff in question, excluding only the parts too small to count. Thus, 'water' and 'sugar' ... are true of each part of the world's water or sugar, down to single molecules but not to atoms" (1960, p. 98). There seems to be a suggestion here of a restriction on the distributive condition, excluding *only* the parts too small to count. But this won't work. There are mereological sums of "parts too small to count", such as the hydrogen and the oxygen in a quantity of water, which are not water but may well be larger than some things which are water. The suggestion might be appropriate for substances that are elemental, but compounds and solutions (homogeneous mixtures—the salt in the sea is not seawater) don't have the same kind of uniformity. An alternative suggestion addressing this latter problem would be to consider specifically spatial parts. Although that wouldn't meet the "parts too small to count" problem, it is of some interest. But first a related problem in connection with what Quine says about mass predicates must be considered.

In recent philosophy, the distributive condition has been supplemented with what Goodman called the collective condition but is more generally known as the *cumulative condition*. Quine says that "mass terms like 'water', 'footwear', and 'red' have the semantical property of referring cumulatively: any sum of parts which are water is water" (1960, p. 91). In terms of the dyadic sum operation, a predicate $\varphi(\pi)$ is cumulative if

$$\varphi(\pi) \wedge \varphi(\rho) \,.\!\supset\, \varphi(\pi \cup \rho).$$

More generally, a predicate, $\varphi(\pi)$, is cumulative iff

$$\exists\pi\,\psi(\pi) \wedge \forall\pi\,(\psi(\pi) \supset \varphi(\pi)) \,.\!\supset\, \varphi(\Sigma\pi\,\psi(\pi)). \tag{4.1}$$

(The existential clause is necessary to ensure the existence of the sum.) Like substance and mixture predicates such as "carbon dioxide" and "aqua regia", phase predicates such as "liquid" and "gas" apparently comply with the condition.

Space and time are sometimes thought to be paradigm examples of mass terms—not, of course, because they are quantities of matter but because they have a non-particulate, formless character that is well characterised by the distributive and cumulative conditions. Only the binary sum and not the general sum is defined for times and spatial regions as introduced in Sects. 1.3 and 1.4. The property of being a spatial region is both distributive—there are no parts of regions other than regions—and cumulative in the sense that the sum of any two regions is a region. Similarly, there are no parts of times (temporal intervals) other than times, and so the property of being a time is distributive. But the question of existence arises for cumulativity for times since sums of arbitrary pairs of times don't exist. With the additional precondition that it applies to times that are connected, ensuring the existence of their sum by MT3 of Sect. 1.3, however, the cumulative condition in terms of the binary sum operation does hold for the property of being a time.

Quantities present a problem for the distributive condition. Quine seems to be suggesting that this be tackled on the basis that the distributive and cumulative conditions are independent, allowing the distributive condition to be denied, and perhaps replaced by some restricted version, while retaining the cumulative condition. But the issue is not so straightforward.

4.3 Relational Predicates

Many predicates of quantities are relational. The sameness of substance predicate "is the same kind of substance", for example, stands between two quantities. Any of the white solid covering much of Greenland is the same substance as the liquid, or rather what constitutes much of the liquid, in the Atlantic. Again, "The water in the kettle is warmer than the surrounding air" ascribes the "warmer than" relation to two quantities. Similarly, we have "weighs more than", "is as massive as", "is more abrasive than", "is harder than", "is more brittle than", "is denser than", "is more opaque than", "has greater reflectance than", "is more rigid than", "is more viscous than", "is more elastic than", and so forth.

In particular, the substance and phase predicates figuring in the last section are in fact dyadic. Substances are fleeting properties of persisting matter. Water is created in the combustion of hydrogen and of hydrocarbons, the metabolism of sugars, what are called condensation reactions in organic chemistry, the neutralisation of acids by alkalis, etc. Water is destroyed by electrolysis, the action of sodium, in photosynthesis, and so forth. And as already mentioned in Sect. 3.1, the picturesque language of coming to be and passing away concerns the possession of properties rather than the existence of their subjects in accordance with Lavoisier's principle of the preservation of mass and the persistence of matter in chemical reactions. Accordingly, the "water" predicate holds of a quantity of matter at one time and perhaps not at another. The predicate "is water" is a two-place relational predicate of the form "π is water at time t". Treated as intervals, times stand, like quantities of matter, in mereological relations of parthood, overlapping and separation, and are subject to the binary mereological operations of sum and product. So the paradigm mass predicates describing substances are not correctly treated as monadic predicates but are relational. Phase predicates are similarly time dependent. Water in a kettle which is boiled becomes steam (gas), and later condenses to liquid again.

The sameness of substance relation is, accordingly, not merely a dyadic predicate "π is the same kind of substance ρ", but a four-place relation with two places for times. Any of the white solid covering much of Greenland is the same substance as the liquid, or rather what constitutes much of the liquid, in the Atlantic. But this holds for quantities of the white solid and the liquid only while they remain water. A quantity which is part of the white solid covering much of Greenland *is* the same substance *now* as a quantity which was liquid in the Atlantic last week *was then*. In the meantime, the liquid evaporated, made its way over the amazon and was incorporated in photosynthesis. So although the former quantity is now the same substance as the latter quantity was, it is no longer the same substance as the latter quantity is. Two quantities of the same kind are not in general indefinitely of the same kind, and the relation to time must be incorporated into a complete representation of the same kind relation, which is a four-place relation "π is the same kind of substance at time t as ρ is at t'". Similar comments apply to other relations mentioned at the beginning of this section. After cooling, a quantity which was warmer than another (was) is no longer warmer than the other quantity was, and so on.

Whilst a quantity of matter may weigh more than another but a part of that first quantity does not, it would seem that any part of a quantity is as warm as another if that first quantity is. Something like the distributive condition might therefore well be applicable to some, if not all, relational predicates relating several entities with a mereological structure. What about the cumulative condition? In order to pursue the issue, the question arises first of how the distributive and cumulative conditions should be generalised to apply to relational predicates. The distributive condition is straightforward. For a dyadic predicate $\varphi(\pi, t)$ applying to a quantity and a time, it generalises to

$$\varphi(\pi, t) \wedge \rho \subseteq \pi \wedge t' \subseteq t \,.\!\supset \varphi(\rho, t'),$$

and predicates of higher arity applying to all kinds of entities with a mereological structure follow the same pattern. But the cumulative condition poses a problem.

Ignoring for simplicity the time-dependency of the "same kind" predicate, a straightforward first shot at generalising the cumulative condition for the dyadic sum operation applied to this predicate reads

(*) If π is the same substance as ρ and π' is the same substance as ρ', then $\pi \cup \pi'$ is the same substance as $\rho \cup \rho'$.

But the antecedent would be true if π and ρ were each water and π' and ρ' each carbon dioxide, whereas the consequent would then be false. Here all π is the same substance as all ρ, and all π' is the same substance as all ρ', but it is clearly not the case that all $\pi \cup \pi'$ is the same substance as all $\rho \cup \rho'$. A similar counterexample is obtained by substituting "same phase" for "same substance" in (*): π and ρ might each be solid and π' and ρ' each gas, but then $\pi \cup \pi'$ and $\rho \cup \rho'$ wouldn't be of the same phase.

Putting it like this may seem to make an appeal to the distributive condition that is questionable at this stage of the argument in view of Quine's objection. But his objection is directed at micro-constituents below the molecular level. These are not at issue in this counterexample because π is a quantity of macroscopic proportions and obviously a part of $\pi \cup \pi'$, and ρ' is similarly a macro-quantity and a part of $\rho \cup \rho'$, but water is not the same substance as carbon dioxide and so π is not the same substance as ρ'. Consequently, it is not the case that all $\pi \cup \pi'$ is the same substance as all $\rho \cup \rho'$, and a restriction meeting Quine's objection would not save the suggested generalisation of the cumulative condition. In any case, were the consequent true, then reapplying (*) to it together with another same substance claim would entail that ever larger and more disparate sums were of the same substance, and eventually all distinctions of substance would be eliminated. The same goes for phase.

It should now be clear that the same substance predicate doesn't relate mixtures of several substances, but has the sense of "same single substance". This is the same substance relation at issue in the proposal that the extension of a substance predicate is determined by what stands in the same substance relation to a paradigm sample of the particular substance found in Putnam (1975, p. 231), except that he imposes a restriction to the same phase. A more general "same stuff", or "same *substances*", predicate would include the relation of being the same mixture, and perhaps (if some philosophers' usage is to be followed) be restricted to certain phases. But that is not at issue here.

Making the time-dependency of substance predicates explicit yields a similar counterexample. Thus,

(†) $\quad Water(\pi, t) \wedge Water(\rho, t') \wedge Ctt' . \supset Water(\pi \cup \rho, t \cup t'),$

is not true in general. The clause Ctt', stating that t and t' are connected (abut or overlap), is necessary to ensure the existence of their sum and thus that the consequent doesn't fail for want of a second term. Suppose that π is a quantity of water at t, and subjected to electrolysis during t', later than but abutting t, when it gradually dissociates into hydrogen and oxygen. Then the sum, $\pi \cup \rho$, wouldn't be water at $t \cup t'$, even if ρ is water during t'. If the consequent of (†) were true, then reapplying (†) to it together with a further claim of a new quantity being water at a time connected with $t \cup t'$, and so on, would entail that ever larger quantities of shifting composition are water for ever larger intervals. Further counterexamples are to be had with "is as warm as", "is warmer than", "has the same pressure as", and the time-dependent phase properties of being solid, liquid and gas.

4.4 Generalising the Cumulative Condition

A satisfactory formulation of the cumulative condition must reduce to the general formulation (4.1) in the special case of monadic predicates and be generalisable in an acceptable way. A solution has been put forward by Roeper (1983) to the effect that (4.1) is equivalent with

$$\forall \pi' \subseteq \pi \, \exists \pi'' \subseteq \pi' \varphi(\pi'') \supset \varphi(\pi). \tag{4.2}$$

Formula (4.2) can be generalised into a suitable cumulative condition for relational predicates which is not subject to the counterexamples which tell against the likes of (*). For a dyadic substance predicate like "water", for example, we have

$$\forall \pi' \subseteq \pi \, \forall t' \subseteq t \, \exists \pi'' \subseteq \pi' \, \exists t'' \subseteq t' \, Water(\pi'', t'') \supset Water(\pi, t).$$

That is, π is water throughout t if some of any part of π during some part of any subinterval of t is all water. This is, perhaps, easier to understand in the contrapositive form: if π is not water throughout t then there is some part π' of π all of whose parts are not water for all subintervals of some part t' of t. Again for the "same substance" predicate, ignoring the time dependence for simplicity as before, we have

$$\forall \pi' \subseteq \pi \, \forall \rho' \subseteq \rho \, \exists \pi'' \subseteq \pi' \, \exists \rho'' \subseteq \rho' \, SameSubst(\pi'', \rho'') \supset SameSubst(\pi, \rho).$$

If π is not the same substance as ρ then there are parts π' of π and ρ' of ρ none of either of which is the same substance as any of the other. With the complete four-place relation, π at t is not the same substance as ρ at t^* implies that there is a part π' of π none of which is ever during some subinterval t' of t the same substance as any part of a part ρ' of ρ for any time during some subinterval t'' of t^*.

The quantification over all parts in (4.2) and the generalised cumulative condition illustrated in these examples smacks very clearly of the distributive condition. In fact, the proof of that part of the equivalence of (4.1) and (4.2) comprising the implication of (4.1) by (4.2), showing that the new condition (4.2) encompasses the cumulative condition, relies on the unrestricted distributive condition. But the proof that the original condition (4.1) implies the new condition (4.2) doesn't. This implication holds simply in virtue of general features of mereology, as shown in the following deduction.

To show that the original cumulative condition implies the new condition, assume the antecedent,

(i) $\forall \pi' \subseteq \pi \, \exists \pi'' \subseteq \pi' \, \varphi(\pi'')$,

of (4.2). Instantiating, $\exists \pi' (\pi' \subseteq \pi \wedge \varphi(\pi'))$, which together with the tautology $\forall \pi'((\pi' \subseteq \pi \wedge \varphi(\pi') \supset (\pi' \subseteq \pi \wedge \varphi(\pi'))$ and the original cumulative condition (4.1) implies $\varphi(\Sigma\pi'(\pi' \subseteq \pi \wedge \varphi(\pi')))$. It remains to show that $\pi = \Sigma\pi'(\pi' \subseteq \pi \wedge \varphi(\pi'))$, so that $\varphi(\pi)$, whence (4.2) follows by the deduction theorem. For reductio, then, assume $\pi \neq \Sigma\pi'(\pi' \subseteq \pi \wedge \varphi(\pi'))$. There are two alternatives. First, suppose $\pi \not\subseteq \Sigma\pi'(\pi' \subseteq \pi \wedge \varphi(\pi'))$. By the strong supplementation principle, there is a σ such that

(ii) $\sigma \subseteq \pi \wedge \sigma \mid \Sigma\pi'(\pi' \subseteq \pi \wedge \varphi(\pi')$.

The first conjunct here together with (i) implies that there is a π'' such that

(iii) $\pi'' \subseteq \sigma \wedge \varphi(\pi'')$.

Then clearly by (ii), $\pi'' \mid \Sigma\pi'(\pi' \subseteq \pi \wedge \varphi(\pi'))$. But by the transitivity of the part relation, the first conjunct of (ii) together with (iii) yield $\pi'' \subseteq \pi \wedge \varphi(\pi'')$. Now if this is true of π'', then clearly π'' is part of the sum of things of which this is true—$\pi'' \subseteq \Sigma\pi'(\pi' \subseteq \pi \wedge \varphi(\pi'))$—contradicting $\pi'' \mid \Sigma\pi'(\pi' \subseteq \pi \wedge \varphi(\pi'))$.

As for the second case, $\Sigma\pi'(\pi' \subseteq \pi \wedge \varphi(\pi')) \not\subseteq \pi$, this implies by the strong supplementation principle that for some π'', $\pi'' \subseteq \Sigma\pi'(\pi' \subseteq \pi \wedge \varphi(\pi')) \wedge \pi'' \mid \pi$. Clearly, the first conjunct implies $\pi'' \subseteq \pi$, contradicting the second conjunct. Both alternatives are contradictory and $\pi = \Sigma\pi'(\pi' \subseteq \pi \wedge \varphi(\pi'))$, so that $\varphi(\pi)$. Q.E.D.

To demonstrate the converse, the consequent of (4.1), $\varphi(\Sigma\pi \, \psi(\pi))$, is deduced from its antecedents and (4.2) with the help of the distributive condition. Assume

(i) $\exists \pi \, \psi(\pi)$
(ii) $\forall \pi \, (\psi(\pi) \supset \varphi(\pi))$

and put $\pi = \Sigma\pi' \, \psi(\pi')$, which exists since by (i) there are ψ-ers. Now by general mereology, anything which is a part of the sum of ψ-ers overlaps some ψ-er; i.e.

(iii) $\forall \pi' \subseteq \pi \, \exists \sigma \subseteq \pi' \, (\psi(\sigma) \wedge \pi' \circ \sigma)$.

But by the definition of overlap, the conjunct $\pi' \circ \sigma$ here implies that there is a part, π'', common to π' and σ, and by the distributive condition (4.1), the first conjunct then implies $\psi(\pi'')$. Accordingly,

(iv) $\forall\pi' \subseteq \pi\, \exists\pi'' \subseteq \pi'\, \psi(\pi'')$.

And so by (ii),

(v) $\forall\pi' \subseteq \pi\, \exists\pi'' \subseteq \pi'\, \varphi(\pi'')$,

and then by (4.2)

(vi) $\varphi(\pi)$,

which is just $\varphi(\Sigma\pi\ \psi(\pi))$, as required. The unrestricted distributive condition was called upon when moving from (iii) to (iv) in this straightforward argument.

Thus the cumulative condition in the form (4.2) is not independent of the distributive condition, which therefore cannot be simply denied while maintaining the cumulative condition in this form for monadic predicates. Clearly, someone who questions the distributive condition for a monadic predicate might question the equivalence. But there is no denying (4.2), which is implied by (4.1) on the strength of purely mereological principles, while accepting cumulativity in its original form. And if Quine's comments on water are to have any bearing, they would have to concern the distributive and cumulative conditions for the dyadic water predicate. The cumulative condition for relational predicates doesn't rely on a corresponding proof of equivalence calling on a relational distributive condition, but it does involve quantification over all parts just as the distributive condition does. In that case, the antecedent of the cumulative condition for "water" is false on Quine's view for any quantity π because it is either some other substance(s) or some parts of it at some time have parts which are all not water throughout some part of this time. The cumulative condition would then be trivially true, and not the substantive truth Quine presumably took it to be. It would seem that the conception of continuous matter expressed by the use of mass predicates implicitly presupposes that they apply to a domain of macroscopic entities none of whose parts are too small to count. Although in the case of times the assumption of infinite divisibility, axiom MT10 of Sect. 1.3, would take us beyond any lower limit of macroscopic duration, it is not easy to see how the assumption itself can be faulted. But quantities, for which neither infinite divisibility nor mereological atomism are assumed, present more of a problem. This is taken up in Chap. 7.

4.5 Spatial Parts

Aristotle rejected atomism, but that is not itself sufficient to save the distributive condition. He did develop a conception of mixts (with no distinction corresponding to that between compounds and solutions) based on the distributive condition, however, with the feature that the original substances from which new mixts are generated are not actually present in these mixts. In particular, the elements are not present in their compounds. Although modern macroscopic theory doesn't follow this analysis, but allows that several independent substances occur in a mixture—

even in the homogenous phases of a mixture that might itself be heterogeneous, the relation between elements in the isolated and combined state retains something of an Aristotelian perplexity. The Stoic view, or at least that aspect of it allowing that several substances are jointly present in a homogenous mixture, may seem more familiar. Spatial parts, on this view, are distinguished from the more general notion of a part, and homogeneity is defined as uniformity over the spatial parts, allowing distinct substances to be distinct proper parts occupying the same place. These ancient theories of substance and mixture are discussed in more detail in the next chapter. But the formulation of distributive and cumulative conditions by restricting reference to parts to spatial part is conveniently developed in this chapter.

Spatial parts of a quantity are not merely parts occupying a part of the region occupied by that quantity. Once cooccupancy is allowed, there is the possibility that there are several distinct parts of the original quantity each occupying a given subregion. It will be possible to talk about parts of spatial parts which are not necessarily spatial parts. But spatial parts are specifically understood to be those *exhausting* the matter in a subregion of that region occupied by the whole quantity. To express this, a predicate, $Exh(\pi, t)$, to the effect that π exhausts all the matter in the region it occupies at t and more briefly read "π is exhaustive at t", is introduced in accordance with

$$Exh(\pi, t) \;\equiv\; \pi \;=\; \Sigma\rho\,(Occ(\pi, t) = Occ(\rho, t)).$$

(The occupies relation $Occ(\pi, p, t)$ is interpreted so that the region, p, occupied by any quantity, π, at a time t is unique, being the sum of all the regions occupied by π at subintervals of t in accordance with the accumulative interpretation discussed earlier, Sect. 2.4.) A dyadic predicate $\varphi(\pi, t)$ is then said to be *spatially distributive* iff

$$\varphi(\pi, t) \;\wedge\; t' \subseteq t \;\wedge\; Occ(\rho, t') \subseteq Occ(\pi, t) \;\wedge\; Exh(\pi, t) \;\wedge\; Exh(\rho, t') \;.\!\supset\; \varphi(\rho, t').$$

This cannot be directly compared with the unrestricted distributivity condition. But under the following assumption of conformity between space, time and matter:

$$t' \subseteq t \;\wedge\; Exh(\pi, t) \;\wedge\; Exh(\rho, t') \;.\!\supset\!.\; Occ(\rho, t') \subseteq Occ(\pi, t) \equiv \rho \subseteq \pi,$$

the spatial distributivity condition simplifies to:

$$\varphi(\pi, t) \;\wedge\; t' \subseteq t \;\wedge\; \rho \subseteq \pi \;\wedge\; Exh(\pi, t) \;\wedge\; Exh(\rho, t') \;.\!\supset\; \varphi(\rho, t').$$

Now it is clear that the antecedent of this condition is stronger than that of the unrestricted distributivity condition for a dyadic predicate $\varphi(\pi, t)$ given in the last section with the additional conjunct $Exh(\pi, t) \wedge Exh(\rho, t')$. It is, perhaps, to be expected that unrestricted distributivity entails but is not entailed by spatial distributivity. This depends, however, on the uniformity assumption.

Spatial parthood between quantities is not simply a matter of parthood between regions because the occupies relation that fixes the regions occupied by the quantities brings in the temporal factor. A spatio-temporal parthood relation of ρ being a part at a subinterval t' of a quantity π at an interval of time t, written $(\rho, t') \sqsubseteq (\pi, t)$, is naturally defined by

$$(\rho, t') \sqsubseteq (\pi, t) \;\equiv .\; t' \subseteq t \;\wedge\; Exh(\pi, t) \;\wedge\; Exh(\rho, t') \;\wedge\; Occ(\rho, t') \subseteq Occ(\pi, t).$$

By the conformity condition, the right-hand side, and so the ostensively time-dependent spatio-temporal parthood relation, implies $\rho \subseteq \pi$. But this is an implication and not an equivalence, so the time dependence has not dropped out of account. With this relation, the spatial distributivity condition can be more compactly expressed as

$$\varphi(\pi, t) \;\wedge\; (\rho, t') \sqsubseteq (\pi, t) \;.\!\supset\; \varphi(\rho, t').$$

In order to compare a corresponding spatial cumulative condition with an analogous restriction to spatial parts, the unrestricted cumulative condition for a dyadic predicate $\varphi(\pi, t)$ is written out explicitly without the restricted quantifiers:

$$\forall \pi' \forall t' (\pi' \subseteq \pi \;\wedge\; t' \subseteq t \;.\!\supset\; \exists \pi'' \exists t'' (\pi'' \subseteq \pi' \;\wedge\; t'' \subseteq t' \;\wedge\; \varphi(\pi'', t''))) \;\supset\; \varphi(\pi, t).$$

Introducing restrictions for the purpose of formulating a spatial cumulative condition analogous to the way the distributive condition was restricted to spatial parts would take the form

$$\forall \pi' \forall t' (Occ(\pi', t') \subseteq Occ(\pi, t) \;\wedge\; t' \subseteq t \;\wedge\; Exh(\pi, t) \;\wedge\; Exh(\pi', t') \;.\!\supset$$

$$\exists \pi'' \exists t'' (Occ(\pi'', t'') \subseteq Occ(\pi', t') \;\wedge\; t'' \subseteq t' \;\wedge\; Exh(\pi'', t'') \;\wedge\; \varphi(\pi'', t'')))$$

$$\supset\; \varphi(\pi, t).$$

Compared with the unrestricted cumulative condition, additional conditions have been introduced both in the antecedent's antecedent and its consequent, and neither the unrestricted cumulative condition nor its spatial-parts variant is stronger than the other. Perhaps a more straightforward entailment, as with the distributivity condition, might have been expected, but this is not to be. With the notion of spatio-temporal parthood in conjunction with the conformity condition, the restricted cumulative condition can be somewhat more compactly expressed by

$$\forall \pi' \forall t' ((\pi', t') \sqsubseteq (\pi, t) \;\supset\; \exists \pi'' \exists t'' ((\pi'', t'') \sqsubseteq (\pi', t') \;\wedge\; \varphi(\pi'', t''))) \;\supset\; \varphi(\pi, t).$$

4.6 Is Cooccupancy Really Possible?

It was suggested that the Stoic view is more in line with modern views of compounds and solutions than the Aristotelian view. But is the fundamental presupposition of the Stoic view, that matter can occupy the same place at the same time, really feasible? Note that what is at issue is the cooccupancy of quantities. Coincidence was taken to be a criterion of identity for individuals in Sect. 3.4. Philosophers have certainly taken sides with Aristotle and balked at the idea of cooccupancy, although without necessarily attending to the distinction between quantities and individuals. It is claimed, for example, that "impenetrability ... implicitly defined by the principle that no two things can be in the same place at the same time ... is entailed by the occupancy of space" (Quinton 1964, pp. 342–3). Aristotle evidently regarded cooccupancy as so self-evidently impossible that reduction to which was sufficient to conclude a reductio ad adsurdum argument. Arguments backing up this stance are thin on the ground. Wiggins (1968, p. 94) claims that distinct objects of the same general kind in the same place cannot be distinguished because "no volume or area of space can be qualified simultaneously by distinct predicates in any range (color, shape, texture and so forth)". But a gaseous solution such as the air exhibits the different partial pressures of each component substance. The various component substance in a mixture may well have different concentrations (mole fractions) and degrees of ionisation, and thermodynamics ascribes to each substance its own chemical potential. An unusual case, by contrast, is the simultaneous exhibition of both a positive and a negative absolute temperature by one and the same lithium fluoride crystal (Zemansky and Dittman 1981, pp. 505–14).

What does science have to say on the subject? In thermodynamics, the energy of a multi-component system subject to work in the form of change in volume is specified as a function $U(S, V, N_1, \ldots, N_r)$ of the entropy, S, the relative amount of each component substance, N_i, and the volume, V. Although the amounts of the several kinds of component substances are specified, only one term for the volume, that of the whole system, appears. This looks very much like cooccupancy. It was certainly the view of Duhem, who studied the matter in detail (Needham 2002, pp. 705–7). Tisza (1977, p. 128) comments that the unusual systems involved in Ramsey's discussion of negative absolute temperature comprise a collection of nuclear spins thermally isolated from the lattice "occupying the same region of space". Remarking that "two isolated systems (spin and lattice) occupying the same region of space is in conflict with [a definition of his theory of macroscopic thermodynamic equilibrium]", he acknowledges that "our basic theory is restricted to a narrower class of systems, in order to yield a larger number of theorems. However, the procedure is consistent with the use of different definitions to suit specialized situations". And modern textbook writers find it natural to speak of "subsystems which occupy the same region of space" (Bowley and Sánchez 1999, p. 189) .

It may seem natural to construe the region occupied by a macroscopic quantity as the sum of the regions occupied by its microconstituents. But we should be wary of imposing conceptions appropriate at the macroscopic level on the microlevel, where matter might well have entirely different kinds of properties. This was clearly appreciated by Maxwell in his speculations about the microrealm. Although taking a view of the macroscopic level at odds with that just outlined, he saw no necessity in applying it at the microlevel: "The doctrines that all matter is extended, and that no two portions of matter can coincide in the same place, being deductions from our experiments with bodies sensible to us, have no application to the theory of molecules" (1867, p. 55). The principles that matter is extended and that quantities of matter with no parts in common occupy regions with no parts in common would each seem to imply that such quantities occupy finite regions. But as already noted these principles don't, on the modern view, apply at the microlevel where there is no clear boundary confining a molecule's electrons. Envelopes confining molecules are conveniently drawn so as to enclose regions where electrons have a 95% probability of being localised. As the electrons move about, on one view of the matter, creating momentary London forces between nearby molecules, simultaneous overlap is precluded by the Pauli exclusion principle. Suppose, for the sake of argument, that molecules are considered to be confined by an envelope. Even if such individual molecules are, for whatever reason, held to be momentarily impenetrable, they will sweep out regions of space in any interval of time due to their considerable translational, rotational and (if sufficiently warm) vibrational motion. For any sufficiently short time when two particular neighbouring molecules don't spatially overlap, it is unlikely that no other neighbouring pairs happen similarly not to overlap. And for intervals of macroscopic order—say of at least one millionth of a second—regions swept out will overlap with many other regions similarly swept out. Accordingly, even if cooccupancy is a problem for localised microparticles with well-defined boundaries at extremely short times, this doesn't translate into a problem for macroscopic times.

Chapter 5
The Ancients' Ideas of Substance

5.1 Doing Without Atoms

Do the distributive and cumulative conditions or their variants restricted to spatial parts reflect important features of significant properties of matter? We will see how they bear on a major division in the physically significant properties of matter between those describing phase and those describing substance. The importance of this distinction became evident in the course of the nineteenth century and it was clearly formulated in the early development of physical chemistry towards the end of the century in the wake of Willard Gibbs' groundbreaking work in the application of thermodynamics to chemistry. Something of the historical origins of these notions and the reasons for their development are traced in this and the next chapter, beginning with two important lines of thought that go back to ancient theories of mixture, namely Aristotle's and the Stoics'. The bearing of more modern understanding of atomic and molecular structure on the main threads of this discussion are taken up in Chap. 9.

The ancients disputed whether matter comprises a single or several substances, but by Aristotle's time several different substances were distinguished and the distinctions of substance were recognised as a major division of kinds of matter. This raised questions about how to determine whether a given quantity of matter comprises a single or several substances, and how different instances of a substance can be recognised as being of the same substance when we come across them on different occasions. Aristotle's own contributions to the subject have sometimes been belittled as more of a hindrance than a help along the road to modern conceptions of chemical substances, largely because he was averse to atomism. Because of the importance of atomism in modern chemistry, the thought seems to be, such antiatomistic theory construction might have something to say about "common sense, or the untutored opinion of children", but must surely be irrelevant to science. As Williams would have it,

P. Needham, *Macroscopic Metaphysics*, Synthese Library 390,
https://doi.org/10.1007/978-3-319-70999-4_5

> the modern theory is alien to Aristotle's thinking on the subject; for in chemical combination the atoms of the combining substances remain intact and change only in respect of their relation to each other. (Williams 2000, p. 142)

But this Democratian vision of atoms certainly isn't the basis of modern chemical theories of combination, which accords better with Aristotle's critical stance. As we will see, he couldn't understand how there could be unchanging atoms unsusceptible to the influence of other matter, and certainly not how combination to form new substances could be explained in terms of such things. Aristotle's positive theory stands more closely to modern macroscopic conceptions of substance than such ancient ideas of what atoms are do to modern microscopic theories of matter. It is true that Aristotelian ideas about the concept of substance had to be overcome in the development of the modern conception of chemical substance. But they played a role in the empirically significant development of chemistry, and were ideas of which even Lavoisier, as we will see, couldn't entirely divest himself.

Lavoisier, one of the founding fathers of modern chemistry, also had no time for atomic speculations, and well-informed scepticism about the value of atomic speculations continued throughout the nineteenth century (Needham 2004a,b; Chalmers 2009). Aristotle seems to have been the first to make a clear distinction between elements and compounds capturing something of the modern conception, evidently anticipating Lavoisier's notion of an element, and there are echoes of his quandary about the fate of elements in compounds that linger on in recent thought (e.g. Paneth 1931), as we will see in Chap. 9.

Duhem (1902), who was in a better position to appreciate the bearing of Aristotelian ideas on modern conceptions of matter than many contemporary commentators, distinguished two main lines of thought about the nature of matter, the atomic and peripatetic traditions. Strangely, there is no mention of the Stoics in this book, although they do appear later in his writings where the distinctive feature of their view of mixture, the cooccupancy of the region occupied by a blend by its component substances, is incorporated into his version of the Peripatetic view of mixts (Needham 2002). Here, the Aristotelian and Stoic conceptions of mixture will be distinguished as two traditions which have severe limitations but have contributed important ideas to the modern conception of substance.

Atomism hasn't made any essential contribution to science much before the turn of the twentieth century. Whilst there has been a longstanding tradition of speculation about the atomic constitution of matter, its contribution to empirical science has been negligible until the time of Perrin's investigations of Brownian motion. Moreover, the best of the traditional arguments against atomism were concerned with the incoherence of the notion, and these were overcome with the realisation that a non-classical theory was required to capture the systematic properties and behaviour of microsystems. The classical conceptions of the atom are paradoxical and beyond redemption. There is an atomic tradition going back to antiquity, but unlike the Peripatetic and Stoic traditions, it is completely severed from the new atomic tradition established in the twentieth century. Chalmers (2009) offers an account and evaluation of the historical tradition of atomism that is not pursued here.

5.2 Aristotle's Conception of Substances as Homogeneous

After a review of earlier work, the development of Aristotle's own view in *De Generatione et Corruptione* (henceforth *DG*) proceeds by first introducing the notion of mixture and with that, the general notion of substance, in book I. His word for what is often translated as "compound"[1] describes the result of a mixing process. But he made no distinction between what, on the basis of the criterion provided by the law of definite proportions early in the nineteenth century, came to be called compounds and solutions. I follow Duhem's use of the antiquated medieval term "mixt" as a technical device for referring to the Aristotelian notion without misleading suggestions for the modern reader.[2]

Aristotle distinguished elements and mixts, but he doesn't assume the distinction from the outset. The concept of an elementary substance is introduced as his discussion develops and is not presupposed in his notion of a mixt. Elements are understood on the basis of the requirements laid down for mixt formation, and defined in terms of characteristics conferring certain powers and distinguished as element-defining characteristics in virtue of being extreme degrees of certain determinables, limiting the number of elements to four. It is a straightforward consequence of the discussion up to this point that no part of a mixture exhibits element-defining characteristics, and thus that there are no elements in mixtures. Aristotle insists that they are potentially recoverable from a mixture, but that is a different matter. It comes as quite a surprise towards the end of book II to see the elements referred to as simple and to be told that each mixt contains all four. The exposition here follows the order of presentation of ideas in *DG*.

His starting point is the observation that there are many substances which are transformed into other substances as they interact. The general problem Aristotle addressed first was how to determine whether a given quantity of matter comprises one substance or many. The key to the question, he thought, was homogeneity, which he formulates in terms of the relation of parthood, the relata of which I call quantities. Aristotle clearly states that all substances (elements and mixts) are homogeneous, or as he puts it, homoeomerous—comprise like parts: "if combination has taken place, the compound must be uniform—any part of such a compound is the same as the whole, just as any part of water is water" (*DG* I.10, 328a10f.). This is just homogeneity because Aristotle simply took it to be evident that parthood is spatial parthood—a point the Stoics were to dispute.

[1]As it is in the Barnes complete edition of Aristotle's works, from which all translations are taken unless otherwise indicated.

[2]Williams speaks of "the modern distinction between chemical combination and mechanical mixture" (2000, p. 142). Although Dalton seems to have had some such notion of a mechanical mixture in mind when contrasting the air as a homogeneous solution with a genuine compound such as water or cupric oxide, the contrast is too stark. Later in the same century the idea of a purely mechanical mixture, termed an ideal solution, was characterised as a mixture arising entirely due to the entropy of mixing. But it is an idealisation, with very few mixtures approaching ideal solution behaviour.

He doesn't give such a nice crisp statement of the converse claim, that homogeneous matter comprises a single substance. But he seems to take it for granted that what he evidently took to be a quantity of homogeneous matter, e.g. a quantity of bone, is a single substance, and the different homogeneous parts of a heterogeneous quantity of matter, e.g. the flesh, blood and bone in an animal body, are different substances. Thus, when encountering diamond in rock, oil in water and smoke or clouds in air, in each of these cases the first claim tells us that there is more than one substance and the second that there are two. Even when an oil and water mixture is shaken and the oil doesn't form a connected body, the oil droplets naturally coalesce when they come into contact, and when left would eventually form a single uniform layer. Proper mixing or mixt formation, then, is a process involving two or more substances being brought together and resulting in homogeneous matter, which by his criterion is a single substance—a new substance arising from the reaction of the initial ingredients, different in kind from either of them.

The idea that heterogeneity marks a distinction of substance also finds expression in his introduction of a special kind of "mixing" process to account for changes like that of water from the liquid to the gas. This kind of mixing is an "overwhelming" process in which one ingredient is present in such a large excess that it isn't itself changed, but completely converts the other ingredient into the same kind as itself. It is introduced in *DG* I.10 alongside proper mixing (described below), and is later applied in *DG* II.4 (after the notion of an element had been introduced) as the mechanism for what he regards as the transmutation of elements, when, for example, water evaporates and becomes what he regarded as the distinct substance, air. So changes known to modern science as phase changes are treated by Aristotle as transmutations of substance. Something like the Aristotelian conception of substances remains in modern chemistry in the use of kind terms like "quartz", which describes the substance silicon dioxide in a particular solid phase, or "diamond", which describes the substance carbon in a particular solid phase, distinct from that exhibited by graphite. The term "water" is still often used in this phase-bound sense, which is defined in the *Shorter Oxford English Dictionary* as "the liquid . . . which forms the main constituent of seas, lakes, rivers and rain . . . ". In this sense, the content of lakes changes to a different substance, ice, when they freeze, and it is not true that all H_2O is water.

Aristotle broaches the notion of an element in the preliminary discussion of his predecessors, when he criticises Anaxagoras for taking homogeneity to be a criterion characterising elements rather than substances in general because this "misapprehends the meaning of element" and leaves no room for the distinction between element and mixt:

> We begin with the view of Anaxagoras that all the homoeomerous bodies are elements. Any one who adopts this view misapprehends the meaning of element. Observation shows that even mixed bodies are often divisible into homoeomerous parts; examples are flesh, bone, wood, and stone. Since then the composite cannot be an element, not every homoeomerous body can be an element; only, as we said before, that which is not divisible into bodies different in form. But even taking 'element' as they do, they need not assert an infinity of elements, since the hypothesis of a finite number will give identical results. Indeed even

> two or three such bodies serve the purpose as well, as Empedocles' attempt shows. . . . it would be better to assume a finite number of principles. They should, in fact, be as few as possible, consistently with proving what has to be proved. . . . Again, if body is distinguished from body by the appropriate qualitative difference, and there is a limit to the number of differences (for the difference lies in qualities apprehended by sense, which are in fact finite in number, though this requires proof), then manifestly there is necessarily a limit to the number of elements. (*De Caelo*, III.4, 302b12–303a3)

Aristotle wants a distinction between elements and substances resulting from the combination of elements when they are brought together in a mixt. This would give point to the notion of elements, which should certainly be finite in number and preferably few. Compounds or mixts, as the names suggest, should be understood as formed from the elements.

The atomists thought that a mixt is just a juxtaposition of the original ingredients, which whether observable or not would persist with their original properties. Aristotle thought this made it a mystery how genuinely new substances with new properties could result from mixing. On his view the original ingredients don't persist in a genuine, homogeneous, mixt. Recovery of the original ingredients entails losing the properties of the mixt.

Aristotle's requirements impose two conditions on elements: they are the things of which non-elemental substances must be considered to be ultimately made, and they are things which have the powers to interact and undergo the changes resulting in mixts with new properties. These conditions are explicated in different places in Aristotle's texts, and both receive formulations which have the appearance of definitions. But whatever position is taken on what Aristotle intended as definitions, there is a substantial problem that was addressed by Lavoisier, who offered a definition of element much like one of the two at issue here and criticised Aristotle for imposing constraints on the nature of elements by the other condition which should rather be open to empirical investigation. Although this criticism was important in the development of science, it isn't an obstacle to giving an exposition of Aristotle's views which highlights the general features that have been important in the development of the concept of chemical substance.

5.3 Two Kinds of Mixing Process

The conception of the elements presented in the first chapters of book II of *DG* is developed in response to the requirements imposed by the understanding of mixing processes and the notion of mixt developed in book I. Quantities of different substances brought into contact will affect one another in accordance with their capacities and susceptibilities. The atomists apparently denied this, but Aristotle couldn't see how atoms could be coherently described without ascribing to them properties which would prevent them persisting unchanged:

> it is impossible that they [the indivisibles] should not be affected by one another: the slightly hot indivisible, e.g., will suffer action from one which far exceeds it in heat. Again, if any

> indivisible is hard, there must also be one which is soft; but the soft derives its very name from the fact that it suffers a certain action—for soft is that which yields to pressure. (*DG* I.8, 326a11–15)

At all events, matter which does enter into mixing processes does so in virtue of capacities and susceptibilities for interaction with other matter, and for Aristotle this comes down to the interplay of contraries. Contrary properties correspond to different degrees over a range of some determinable, and two quantities of matter bearing contrary properties might be expected to interact with the net result that some intermediate degree of the same determinable is realised by the entire matter comprising the mereological sum of the two quantities. If all such contrary pairs of the two initial quantities are similarly neutralised, perhaps all qualitative differences between the two are obliterated with the result that the mereological sum of the two quantities will be a homogeneous quantity in which the original two quantities are not marked off from one another by any physical discontinuity. Where this happens, and the result is a homogeneous mixt, we have mixing of the first kind.

Mixt formation is not the only possible result of the interplay of contraries that Aristotle envisages. It requires that the quantities of interacting matter are of "comparable" amounts, and is contrasted with another kind of mixing process that occurs when one of the interacting quantities is of an overwhelmingly larger amount than the other. When mixing of this second kind occurs, the contraries of the larger quantity so overwhelm those of the other that the result is a mereological sum of the two quantities all of which bears the contraries initially born by the larger quantity without any remission of degree. This is the kind of mixing process that Aristotle later appeals to in chapter 4 of book II of *DG* to explain what he regards as processes involving the transmutation of the elements. Although this is central to Aristotle's conception of the elements, it has no direct bearing on the issues which are taken up in the following discussion. It will therefore not be necessary to dwell on the obvious problems about which measurable magnitude determines the amounts at issue and how to draw a boundary between "comparable" and "overwhelming" amounts.

The discussion proceeds in book II by considering the interaction of contraries in more detail and advancing the claim that all the causally effective features of substances bearing on how they interact are reducible to degrees of the primary determinables warmth and humidity. This, at any rate, is fairly clearly the conclusion of the discussion in chapter 2 of book II, although the argument is very sketchy and leaves it unclear by what criterion all relevant properties are deemed to have been taken into account. But heaviness and lightness—notions which play a role in other aspects of Aristotle's physics—are excluded from these reductionist considerations since "Things are not called heavy and light because they act upon, or suffer action from, other things" (*DG* II.2, 329b22f.). It is also clear that these primary determinables are taken to be closely associated with phase properties, so that being wet is associated with being fluid (liquid or gas, in virtue of which a body adopts the shape of a container) and dry with solid (having a shape of its own). Having a volume of its own, which distinguishes liquids from gases, calls for some distinguishing feature other than wetness which they share. Aristotle's view, then, has it that degrees of warmth and humidity are primary in the sense of not being reducible to any other determinables, whilst other determinables are reducible to them. It is not difficult to understand that bodies with different degrees of warmth should affect one another when brought into contact, the warmer one becoming cooler whilst the cooler becomes warmer until a uniform state of equilibrium is reached, and a wetter and a dryer body interact analogously to attain a shared intermediate degree of humidity.

Aristotle goes beyond the assumption of the existence of mixed substances, characterised by degrees of the primary determinables, by making the further assumption that the primary determinables of warmth and humidity are bounded above and below, i.e. that there are maximal and minimal degrees of these primary determinables, what he later calls "contrary extremes" (*DG* II.8, 335a8) of warmth and humidity. A quantity of matter is said to be cold if it has the minimal degree of warmth, hot if it has the maximal degree, dry if it has the minimal degree of humidity and wet if it has the maximal degree. This postulate complements the postulation of mixts, generated by mixing some other substances, by ensuring that there are substances from which all mixts can be considered to have been made. (The existential quantifier here stands outside the scope of the universal quantifier.) There are no degrees of the primary determinables outside these limits which could characterise substances which would generate these ingredients from which all other substances are in fact made upon mixing. Aristotle is then in a position to introduce elements as *substances characterised by extreme degrees of the primary determinables* (what I will call the first definition of "element"). On the assumption that warmth and humidity are the only primary determinables, and since the extreme degrees of a given primary determinable are clearly mutually exclusive, it follows that there are just four elements corresponding to the four compatible combinations of the extremes. Aristotle deems water to be what is cold and wet, earth to be what is cold an dry, air to be what is hot and wet and fire to be what is hot and dry (*DG* II.3, 330a30–330b5).

It is natural to understand these primary determinables and the corresponding extremes as distributive, as this term is understood in the general theory of mass predicates. Thus, if a quantity of matter has one of the features in question, so does every part of that quantity. Parts of a quantity of different degrees of warmth, for example, would naturally come to equilibrium at an intermediate degree if the parts were in contact. A quantity might be disconnected, of course. But if the parts of a quantity are not of the same degree of warmth, it is natural to say that the quantity doesn't have a degree of warmth, just as with the modern notion of temperature, a body whose parts are not at the same temperature doesn't have a temperature. This would accord, in view of the definitions of each of the elements just given, with what Aristotle says about a mixt being uniform in the sense that the parts are like the whole in the passage quoted from *DG* I.10 at the beginning of Sect. 5.2. And among the properties uniformly distributed over a mixt will be the determinate degree of the determinable corresponding to the contraries which attain a common intermediate degree in the mixt.

It follows that there are no elements in mixts. When elements come together and form a mixt as a result of a mixing process of the first kind, the quantities of matter which are of these element kinds lose the features which characterise elements. There is an interaction of contraries, and intermediate degrees of the primary determinables are attained, distinct from any extreme degrees. Since a mixt is, as Aristotle says, uniform, so that all the parts have the same characteristic features as the whole, then all the parts have these intermediate degrees of the primary determinables. There are no parts with the extreme degrees, and thus no parts which are elements.

5.4 Simplicity, Actuality and Potentiality

Elements are not distinguished from mixts in virtue of all their parts being of the same kind, i.e. in terms of actual composition. All substances are uniform in this sense, and mixts are not distinguished as substances containing other substances which don't in turn contain any other substances. This would seem to preclude the notion of an element as a distinctively simple substance, at least on one contemporary understanding of the notion. But when Aristotle goes on in *DG* II.3 to agree with "all who make the simple bodies elements" (330b7), he has in effect given "simple" the sense of "not generated by a process of mixing other substances". This is presumably how his speaking of elements as simple later on in *DG* II.8, with the claim that "All the compound bodies ... are composed of all the simple bodies" (334b31), is to be understood. It gives what might be called the synthetic sense of simplicity.

Distinct mixts must be characterised by distinct degrees of either warmth or humidity. If they are all to be conceived as being derived from mixing all four elements, it might seem natural to consider the differences in characteristic degrees of either warmth and humidity to derive from different proportions of the elements. There are occasional references to elemental ratios, but this is problematic. Proportions of elements in compounds have been understood since Lavoisier on the basis of the concept of mass. But this wasn't available to Aristotle, whose conception of proportion must have some other basis. A line of speculation on this issue is developed in Needham (2009a) but not pursued here.

Aristotle addresses the notion of simplicity differently in *De caelo*. How the account he offers in terms of possessing "a principle of movement in their own nature" (268b28) squares with views offered elsewhere is not clear. But later in *De caelo* he presents what, at least on the face of it, seems to be yet another conception of simplicity, although more clearly related to the considerations central to the *DG* discussion and might be called an analytic conception (the second definition) of elements.

> An element, we take it, is a body into which other bodies may be analysed, present in them potentially or in actuality (which of these is still disputable), and not itself divisible into bodies different in form. That, or something like it, is what all men in every case mean by element. (*De Caelo* III.3, 302a15ff.)

For consistency, the actuality option is clearly excluded by the considerations at the end of the last section, and the analytic notion of simplicity is a modal concept, whether or not this is what everyone always means by element. This move is anticipated in the discussion of mixing in *DG* I.10, where Aristotle introduces the distinction between actual and potential, albeit with some ambiguity:

> Since, however, some things are potentially while others are actually, the constituents can be in a sense and yet not-be. The compound may be actually other than the constituents from which it has resulted; nevertheless each of them may still be potentially what it was before they were combined, and both of them may survive undestroyed. (*DG* I.10, 327b23ff.)

Saying that the elements "survive undestroyed" certainly seems to undermine the potentiality claim, but that suggestion can't be taken literally. Speaking of separation of the elements also seems to imply a spatial separation of elements that have been present all along. In the same vein the word "mixture", as Aristotle points out at the beginning of *DG* I.10, carries a suggestion of actual presence which I have tried to counter by speaking of the mixing process and the result as a mixt. The Stoics were not just making a trivial grammatical point when they claimed that separation is only possible of the elements which are actually present in a mixt. They disputed Aristotle's substantial thesis that it is possible to subject a mixt to a separation process resulting in the isolation of substances of the same kind as those from which it was originally made by a mixing process. Nevertheless, it is convenient to speak of separation of the original ingredients where the context makes it clear that it is Aristotle's understanding of this as the emergence of new substances that is at issue.

What does survive undestroyed is the matter which bears the different substance properties at different times in the course of the mixing and separation processes, and what are preserved are the modal or dispositional properties when in the state of the mixt. Speaking of quantities of matter as the bearers of these varying properties in this way accords with the conservation of matter interpretations of Weisheipl (1963), Bostock (1995) and more recently Cooper (2004). Cooper opposes his interpretation to that of a certain tradition among commentators going back to Philoponus who interpret Aristotle's claim in *DG* II.8, that "All the compound bodies ... are composed of all the simple bodies" (334b31), to imply a "total interfusion" view according to which "the smallest bit of my flesh is put together from all the simple bodies" (Cooper 2004, p. 325). Against this, Cooper wants to allow the possibility that a small bit of flesh "did not derive from any water, but only from some earth" (p. 325), which he maintains is consistent with the requirement of total uniformity of mixts (the distributive condition). Cooper emphasises that a mixt "results from the mutual interaction of finitely small bits of the separate ingredients acting on one another" (p. 323), but nevertheless acquires the common features characteristic of the mixt because "what was earth, say, now comes to have just the same perceptible characteristics as the fire, and/or the air or the water, that it has been mixed with also come to have" (p. 322).

Cooper seeks to maintain the distinction between the Aristotelian and the Stoic views which the Philoponus interpretation threatens to run together. I would put his point as building on the distinction between a property and what it applies to. The latter I call quantities of matter, which are treated as things bearing different substance properties (being of different kinds of substance) at different times. This provides for distinguishing between what is and isn't preserved when Aristotle says "The compound may be actually other than the constituents from which it has resulted; nevertheless each of them may still be potentially what it was before they were combined, and both of them may survive undestroyed" (*DG* I.10, 323b25–27). What is not destroyed are the quantities of original ingredients which came together to make the mixt, although the original substance kinds cease to apply to these quantities when forming parts of the mixt. Thus, we can speak with Cooper of what was earth and is now a mixt and may later again be an elemental kind, namely a quantity of matter. The quantities of matter themselves are identifiable throughout such transmutations, in accordance with the mereological criterion of identity, and are not interpenetrable on the Aristotelian view.

What seems to me to be an outstanding difficulty with Aristotle's view is his claim that "The constituents, therefore, neither persist actually, as body and white persist; nor are they destroyed (either one of them or both), for their potentiality is preserved" (*DG* I.10, 323b29–31). What drives the process of decomposition in which this potentiality is realised is not so readily understood as the neutralisation of contraries drives mixing. It is unclear what should induce a particular part to become, say, water even if that quantity was water before mixing since, if the mixt is uniform, it is like all the other parts in respect of the relevant features of warmth and humidity. In virtue of what, then, is the specific potentiality to become water preserved? This problem is comparable to the chief difficulty facing the Stoic view, of providing a characterisation of the nature of elements common to the isolated and blended state (discussed in the next section). Elsewhere, however, Aristotle suggests that it must be possible to derive the elements from "any and every part of flesh" (*DG* II.7, 334a34), so that the original quantities don't preserve the potentiality to become the same element, but each part of the mixt has the potentiality to become any of the elements (under the constraint that decomposition of the entire mixt will result in quantities of the four elements standing in the same proportion as did the original ingredients). So specific quantities of elements don't preserve a potentiality to become the same element when in a mixt, just some element.

It does seem reasonable, then, to interpret the Aristotelian view as one based on an assumption of the indestructibility of matter which is reflected in the treatment of quantities in the mereological interpretation offered here and in Cooper's use of expressions like "what was earth ... now comes to have the same ... characteristics as fire". Lavoisier was able to take advantage of Newton's conception of mass as a measure of the amount of matter, which wasn't available to Aristotle, and formulate this principle as the law of the conservation of mass. But the underlying metaphysical principle is the same as Aristotle's. They differed on the question of the permanence of the elements, which is not part of Aristotle's scheme. The Stoics' view was closer to Lavoisier's. They entertained a conception of the permanence of the elements involved in the formation of blends in order to explain the possibility of separation (even if they also allowed for the destruction of elements in fusions).

5.5 The Stoic Alternative to Potential Presence

The elements are not actually present in mixts, according to Aristotle. Mixing is driven by the powers to affect and susceptibilities to be affected of the initial ingredients, which change when these powers have run their course and a new substance is formed. In accordance with the homogeneity requirement, that "if combination has taken place, the compound must be uniform—any part of such a compound is the same as the whole, just as any part of water is water" (*DG* I.10, 328a10f.), every part of a mixt has the same intermediate degrees of the qualities characterising the initial ingredients. There are thus no parts with properties characteristic of the elements, namely the extremes of warmth and humidity, and elements are nowhere to be found in a mixt. But although not actually present, Aristotle says that they are potentially present in the sense that they are potentially recoverable.

As we saw, whereas the neutralisation of the characteristic elemental properties when brought into contact by the attainment of intermediate degrees of warmth and humidity is intuitively understandable, it is difficult to understand what, on Aristotle's view, drives the generation of elements by decomposition of a mixt. Perhaps the introduction of the second, overwhelming, kind of mixing was intended to address this problem. But we might sympathise with the Stoics who sought a view of mixts which made decomposition into elements more readily intelligible.

The Stoics distinguished two kinds of mixt. One results from a mixing process that Chrysippus, according to Alexander of Aphrodisias, called "total fusion with both the substances and their qualities being destroyed together". This, Alexander continues, he "says happens with medical drugs in the joint-destruction of the constituents and the production of some other body from them" (216.14).[3] Fusions are homogeneous, like Aristotle's mixts, but differ from them in being indecomposable. Blends, on the other hand, are also like Aristotelian mixts in being homogeneous, but unlike fusions, their original ingredients are separable because they are actually present in the blend. Stobaeus gives the example of the separation of water from wine: "That the qualities of blended constituents persist in such blendings is quite evident from the fact that they are frequently separated from one another artificially. If one dips an oiled sponge into the wine which has been blended with water, it will separate the water from the wine since the water runs up into the sponge" (Long and Sedley 1987, p. 291).

[3]References in this form are to Todd (1976), by the number of the first line of the paragraph from which the quotation is taken.

The apparent conflict between the requirements of homogeneity and actual presence of the elements in a blend is avoided by rejecting Aristotle's view that two quantities of matter cannot occupy the same region of space at the same time. The Stoics held that spatial uniformity is consistent with the elements composing a compound occupying the same region as the compound and as one another. Alexander tells us that Chrysippus held that "blending in the strict sense of the term" results in "[t]he mutual coextension of some two or even more bodies in their entirety with one another so that each of them preserves their own substance and its qualities in such a mixture . . . ; for it is a peculiarity of bodies that have been blended that they can be separated again from one another, and this only occurs through the blended bodies preserving their own natures in the mixture" (216.14).

The Stoic view of blends would circumvent the Aristotelian conception of the potential presence of elements in mixts based on the intuition that separation is only possible of what is actually present in the mixt. This may sound quite promising, but it faces the problem of characterising the elements in terms of properties exhibited in both isolated and combined states, i.e. to describe their "natures" which are preserved in blends. Aristotle defined the elements by conditions they exhibit in isolation—conditions which nothing in a mixt can satisfy. But this strategy is clearly not available on the Stoic view. The Stoics don't seem to have been able to meet this challenge.

Alexander doesn't say what the Stoic view on elements was. Bréhier (1951, Ch. II, §2) says that Chrysippus subscribed to the common view of the four elements, intertransmutable as for Aristotle, albeit by a different mechanism (pp. 138–9). The Stoic view must differ from the Aristotelian view by requiring that features characterising the elements are exhibited in the blended state—an impossibility on the Aristotelian view in so far as blends are homogeneous—to account for the possibility of separation. And this, as Alexander in effect points out, precludes adopting the Aristotelian definitions of the elements on the Stoic view. For "how could anyone say that there is something actually hot in what is cold?" (224.22). Some ancient sources suggest each element was defined by just one of the four primary qualities Hot (fire), Cold (air), etc. (Bréhier 1951, p. 136). But this doesn't avoid the problem of contraries in the same place. Hahm (1985, pp. 42ff.) suggests that the elements were characterised by differences in density. But this would mean that the elemental properties are not preserved in a blend, since a quantity would occupy a greater volume when part of a blend compared with the isolated state.

Aristotle meets the problem of mixture in so far as he proposes that elements have properties which explain how they interact (effect one another) and produce something with different properties by the straightforward (if naive) device of saying that the resultant matter has degrees of the determinables in question intermediate between those of the interacting elements. Their not being present in the compound doesn't prevent them interacting to generate the compound in this way. But the process of decomposition of a mixt into elements is not as readily understandable as Aristotle's account of mixt formation. And as empirical theories, neither kind of process seems particularly well founded. But on the general issue of principle,

Aristotle has a coherent conception of elements and their relation to substances formed from them, whereas the Stoic theory doesn't seem to have a concept of element appropriate to it.

5.6 Formalisation

The Aristotelian theory lends itself to a formalisation within a mereological framework providing for the introduction of a number of substance properties, $S_1, S_2, \ldots$ by a procedure described below. These are two-place relations, $S_i(\pi, t)$, relating a quantity of matter and a time. The time dependence arises because substances undergo change into other substances via mixing processes. This holds as much for elementary substances as for mixts because on the Aristotelian theory, even the elements undergo transmutation into other elements as well as into mixts. They all satisfy the *distributive condition* of applying to all the parts of whatever they apply to:

$$S_i(\pi, t) \;\wedge\; \rho \subseteq \pi \;\wedge\; t' \subseteq t \;.\!\supset\; S_i(\rho, t'). \tag{5.1}$$

All substances are characterised in terms of the two transitive and reflexive relations, $\geq_w$, of being at least as warm as and $\geq_h$, of being at least as humid as. These relations change from time to time, however, and should therefore be expressed as three-place, time-dependent relations between two quantities and a time, $\geq_w (t)$ and $\geq_h (t)$. Transitivity and reflexivity are accordingly understood to hold for a fixed time, and expressed as follows:

$$\pi \geq_w (t)\rho \;\wedge\; \rho \geq_w (t)\sigma \;.\!\supset \pi \geq_w (t)\sigma. \tag{5.2}$$

$$\pi \geq_w (t)\rho \;\supset \pi \geq_w (t)\pi \tag{*}$$

and similarly for $\geq_h$. Given that times are intervals, (5.2) can be extended to the effect:

$$\pi \geq_w (t)\rho \;\wedge\; \rho \geq_w (t)\sigma \;\wedge\; t' \subseteq t \;.\!\supset \pi \geq_w (t')\sigma. \tag{5.2a}$$

It is natural to assume that these warmth and humidity relations are distributive over time, and also that they hold between all the parts of any quantities between which they hold. In this respect they resemble the intensive character of temperature (and the corresponding relation "at least as warm as") in equilibrium thermodynamics. Accordingly, any quantity as warm (humid) as itself is uniform in the sense of being as warm (humid) as each of its parts, and so homogeneous (because Aristotle thought phase properties somehow reducible to warmth and humidity). From the modern point of view, equilibrium conditions don't hold universally. But Aristotle assumes that substance properties are distributive, and that being a particular kind of substance corresponds to having particular degrees of warmth and humidity uniformly. Perhaps material can be in such a condition that its parts are not of definite substance kinds. But for simplicity I assume this is not the case on the Aristotelian view and that the warmth and humidity relations are distributive, i.e.

$$\pi \geq_w (t)\rho \wedge \pi' \subseteq \pi \wedge \rho' \subseteq \rho \wedge t' \subseteq t \, .\supset \pi' \geq_w (t')\rho', \tag{5.3}$$

and similarly for $\geq_h$. In that case, the restriction of reflexivity can be dropped and (*) replaced by

$$\pi \geq_w (t)\pi. \tag{5.4}$$

The relations $\approx_w$ and $\approx_h$, of being as warm (humid) as, are equivalence relations (for fixed time) when defined by

(D1) $\pi \approx_w (t)\rho \;\equiv .\; \pi \geq_w (t)\rho \wedge \pi \leq_w (t)\rho$

(D2) $\pi \approx_h (t)\rho \;\equiv .\; \pi \geq_h (t)\rho \wedge \pi \leq_h (t)\rho,$

where $\pi \leq_w (t)\rho$ (for "π is at least as cold as ρ") is just the converse, $\rho \geq_w (t)\pi$, of "at least as warm as" and similarly $\leq_h (t)$ is the converse of $\geq_h (t)$. Aristotle's view that particular substances correspond to particular degrees of warmth and humidity is captured by defining the same substance relation *SameSubst* as follows:

(D3) $SameSubst(\pi, \rho, t) \;\equiv\; \pi \approx_E (t)\rho,$

where $\pi \approx_E (t)\rho$, for "π is at equilibrium with ρ at t", is defined as $\pi \approx_w (t)\rho \wedge \pi \approx_h (t)\rho$. From the fact that $\approx_w$ and $\approx_h$ are distributive equivalence relations it follows that for fixed *t*, *SameSubst* is also a distributive equivalence relation. Then the *SingleSubst* relation applies to a quantity of matter, π, which is a single substance throughout some time t, just in case its every part is the same substance:

(D4) $SingleSubst(\pi, t) \;\equiv .\; \forall\rho\,\forall t'(\pi \subseteq \rho \wedge t' \subseteq t \, .\supset SameSubst(\pi, \rho, t')).$

A two-place predicate, S_1, can now be introduced to describe this substance, and this predicate satisfies condition (5.1). A quantity which is not the same substance as a given quantity but is nevertheless itself a single substance is a distinct substance:

(D5) $DistSubst(\pi, \rho, t) \equiv . \sim SameSubst(\pi, \rho, t) \wedge SingleSubst(\pi, t) \wedge$
$SingleSubst(\rho, t).$

With the aid of some additional principles introduced shortly it will be possible to show that D5 implies that π and ρ are separate.

A two-place predicate, S_2, can be introduced to describe a substance distinct from substance S_1, a further two-place predicate, S_3, can be introduced to describe a substance distinct from each substance S_1 and S_2, and so on for as many substances as can be found, all these predicates satisfying condition (5.1). This procedure leaves open the determination of characteristic properties, in addition to the fundamental ones of degree of warmth and humidity, of each of these substances, which can be assembled in lists amounting to an alternative means of establishing distinctness of substance when direct mixing is (producing something with degrees of warmth or humidity different from any of the initial ingredients) inconvenient. It is in this way that the elements are distinguished from substances in general.

The elements are characterised by having extreme degrees of warmth and humidity. To express this idea, we first have to lay it down that warmth and humidity are bounded above and below:

$$\exists \pi \, \forall \rho \, (\rho \geq_w (t) \pi) \; \wedge \; \exists \pi \, \forall \rho \, (\rho \leq_w (t) \pi), \quad (5.5)$$

and similarly for $\geq_h$. (As it stands, this holds for all times, t. Modifications, including the possibility of accommodating the principle in modal terms, will be considered shortly.) The extremal property of being wet (W) can now be defined as being at least as humid as anything:

(D6) $W(\pi, t) \; \equiv \; \forall \rho \, (\pi \geq_h (t) \rho)$,

and similarly, being dry (D) is defined as everything being at least as humid, being hot (H) as being at least as warm as anything and being cold (C) is defined as everything being at least as warm:

(D7) $D(\pi, t) \; \equiv \; \forall \rho \, (\rho \geq_h (t) \pi)$,

(D8) $H(\pi, t) \; \equiv \; \forall \rho \, (\pi \geq_w (t) \rho)$,

(D9) $C(\pi, t) \; \equiv \; \forall \rho \, (\rho \geq_w (t) \pi)$.

The elemental substance property $Water(\pi, t)$ can then be defined as $W(\pi, t) \wedge C(\pi, t)$, and similarly for $Earth(\pi, t)$, $Air(\pi, t)$ and $Fire(\pi, t)$ in accordance with Aristotle's definitions. A mixt is a non-elemental substance with non-extremal degrees of warmth and humidity, i.e. a quantity which, at the time in question, is either strictly warmer than something and strictly colder than something else, or strictly more humid than something and strictly less humid than something else (or both).

Further development is possible once variables $p, q, r, \ldots$ referring to spatial regions and the predicates $Occ(\pi, p, t)$ for "π occupies exactly the region p throughout t" and Apq for "p abuts q" are brought into the picture. A principle common to both Aristotelian and Stoic views is that spatial separation implies separation:

$$Occ(\pi, p, t) \; \wedge \; Occ(\rho, q, t) \; \wedge \; p \mid q \,.\!\supset \pi \mid \rho. \quad (5.6)$$

So if π and ρ overlap, so do the regions they occupy at any time. The fundamental Aristotelian principle of the impossibility of cooccupancy can be formulated for stationary quantities as

$$Occ(\pi, p, t) \; \wedge \; Occ(\rho, q, t) \; \wedge \; \pi \mid \rho \,.\!\supset p \mid q. \quad (5.7a)$$

Allowing that π and ρ might move, sweeping out regions which overlap, a more restricted principle along the lines of

$$Occ(\pi, p, t) \wedge Occ(\rho, q, t) \wedge \pi \mid \rho \,.\!\supset \tag{5.7b}$$
$$\exists p' \subseteq p\, \exists q' \subseteq q\, \exists t' \subseteq t\, (Occ(\pi, p', t') \wedge Occ(\rho, q', t') \wedge p' \mid q')$$

is required to cover the general case.

The question now is how to express Alexander's general principle that different degrees of the same primary determinable of warmth or humidity cannot be instantiated by matter in the same place. A first shot might be

(**) $Occ(\pi, p, t) \wedge Occ(\rho, q, t) \wedge \sim(\pi \approx_w (t)\rho) \,.\!\supset p \mid q,$

and similarly for the $\approx_h$ relation. But this is equivalent to saying that quantities occupying overlapping regions have the same degree of warmth, which is too strong. Even if the parts cooccupying the regions of overlap are as warm as each other, they might have other parts not as warm as one another. A safer formulation would be:

$$Occ(\pi, p, t) \wedge \forall \pi' \subseteq \pi\, \forall t' \subseteq t\, (\pi' \approx_E (t')\pi) \,.\!\supset \tag{5.8}$$
$$\forall \rho\, \forall q \subseteq p\, \forall t' \subseteq t\, (Occ(\rho, q, t') \supset \rho \approx_E (t')\pi.$$

With these postulates, $DistSubst(\pi, \rho, t)$ implies $\pi \mid \rho$ as the following argument sketch shows. By D5, $DistSubst(\pi, \rho, t)$ implies $\sim SameSubst(\pi, \rho, t)$, which in turn implies that π and ρ differ either in warmth or humidity. But since $DistSubst(\pi, \rho, t)$ also implies that π and ρ are each single substances, their warmth and humidity is uniform, i.e. the same for all of their respective parts. Accordingly, each part of π differs from each part of ρ either in warmth or humidity, and by (5.8), occupies different spatial regions. Therefore, π and ρ occupy regions that are separate, and so are themselves separate by (5.6).

With a view to expressing that the result of Aristotelian mixing is the attainment of uniform degrees of warmth and humidity, recall (Sect. 1.4) that two regions are connected if they either abut or overlap: $Cpq \equiv .\ Apq \vee p \circ q$. Two quantities, π and ρ, are then said to be connected throughout t if they occupy connected regions throughout t:

(D10) $Conn(\pi, \rho, t) \equiv \forall t' \subseteq t\, \exists p\, \exists q\, (Occ(\pi, p, t') \wedge Occ(\rho, q, t') \wedge Cpq).$

Two quantities of matter, which might have had different degrees of warmth or humidity before t, will in general come to have the same degrees of warmth and humidity if they remain in contact for a sufficiently long time after coming together at t. The present apparatus, which is not adequate to express this principle, limits us to some such (unrealistic) condition as quantities in contact are at equilibrium:

$$Conn(\pi, \rho, t) \supset .\ \pi \approx_w (t)\rho \wedge \pi \approx_h (t)\rho. \tag{5.9}$$

The common degrees of warmth and humidity attained on contact could be further specified to be intermediate between the original degrees. But note that this latter comparison involves two times, to the general effect that the sum of π and ρ is warmer than π at t than it was at t', and the sum of π and ρ is colder than ρ at t than it was at t', and similarly for the humidity relation. These relations can be expressed as $\pi \cup \rho \geq_w (t', t)\pi \wedge t' < t$, and $\pi \cup \rho \leq_w (t', t)\rho \wedge t' < t$, where $t' < t$ stands for "t' is earlier than t". The original three-place warmer than relation (expressed with a single time variable) can be defined in terms of the new four-place relation by identifying the two times:

(D11) $\pi \leq_w (t)\rho \equiv . \pi \leq_w (t', t)\rho \wedge t' = t.$

Now that it has been recognised that relational comparison can involve several times, we can return to the question of how the upper and lower bounds of the as warm as and as humid as relations, expressed above by (5.5), might be elaborated. As it stands, (5.5) implies that there is always something at least as warm as anything and there is always something as least as cold as anything. This allows that different things exemplify these extremal properties at different times, which is not itself problematic, but it leaves open the question of some sort of uniform standard over time.[4] This idea can be captured by modifying (5.5) to read

$$\exists\pi\, \exists t\, \forall t'\, \forall\rho\, (\rho \geq_w (t', t)\pi) \wedge \exists\pi\, \exists t\, \forall t'\, \forall\rho\, (\rho \leq_w (t', t)\pi), \tag{5.10a}$$

to the effect that there is something which is at some time no warmer than (and no colder than) anything ever is, and similarly for $\geq_h$.

It might be thought that the idea of extremal degrees of the primary determinables is a modal notion—that they are never actually realised, but are only possibilities (capturing, perhaps, the claim expressed by some commentators that the Aristotelian elements are ideal notions, never in fact actually exemplified by any matter). The expression of this idea presents a problem, however. This idea is not captured by merely weakening (5.10a) to

$$\Diamond\, \exists\pi\, \exists t\, \forall t'\, \forall\rho\, (\pi \geq_w (t', t)\rho) \wedge \Diamond\exists\pi\, \exists t\, \forall t'\, \forall\rho\, (\pi \leq_w (t', t)\rho). \tag{5.10b}$$

What we want to say is that there might be something which is at some time as warm as anything ever actually is (or could be). But what (5.10b) in fact says is that it is possible that there is something which is at some time as warm as anything might ever be (as distinct from "as anything ever actually is"). This allows that there isn't actually any upper bound on the as warm as relation, just that there might be. The problem might be approached by indexing occurrences of variables predicated

[4]In modern science, a calibration point on the absolute temperature scale is fixed by reference to a reproducible state of water, when gas, liquid and solid are jointly exhibited by a quantity of the substance at its so called triple point. According to thermodynamics, this occurs at a definite temperature and pressure.

by the "as warm as" predicate, say by attaching superscripts to the variables thus: ${}^1\pi \leq_w (t, t')^2\rho$ and expressing the intended thought by

$$\Diamond^1 \; \exists\pi \, \exists t \, \forall t' \, \forall\rho \, \Delta^2 \, ({}^1\pi \geq_w (t', t)^2\rho) \wedge$$
$$\Diamond^1 \; \exists\pi \, \exists t \, \forall t' \, \forall\rho \, \Delta^2 \, ({}^1\pi \leq_w (t', t)^2\rho), \qquad (5.10c)$$

where Δ^i is an expression of actuality governing the ith position within its scope. Alternatively, a stronger statement can be made by replacing Δ^2 in (5.10c) with an expression of necessity, $\Box^i$, defined as $\sim \Diamond^i \sim$. These expressions of modality are not the usual operators, but predicate modifiers which cannot be iterated. It seems that what is at issue here is what were called modal comparisons at the end of Sect. 3.7, but which cannot be captured by resorting to rigid mereological relations as was done there. The idea is pursued in Chap. 9.

The Stoics' view was closer to Lavoisier's. They entertained a conception of the permanence of the elements involved in the formation of blends in order to explain the possibility of separation (even if they also allowed for the destruction of elements in fusions). The basic premise of the Stoic view is the possibility of cooccupancy, definable as

(D12) $Coocc(\pi, \rho, t) \;\equiv\; \exists p \, (Occ(\pi, p, t) \wedge Occ(\rho, p, t))$.

This relation holds trivially where $\pi = \rho$, but the contention is that it can hold for distinct, indeed separate, π and ρ. This could be used to define a blend along the lines of

(D13) $Blend(\pi, \rho, t) \;\equiv\; (\pi \mid \rho \wedge Coocc(\pi, \rho, t) \wedge DistSubst^*(\pi, \rho, t))$.

if there were an account available (implying $\pi \mid \rho$) of the relation $DistSubst^*(\pi, \rho, t)$ of π and ρ being distinct substances throughout t on the Stoic view. As we have seen, however, it is unclear by what criterion this latter relationship is established on the Stoic view. Heterogeneity was Aristotle's criterion, but that clearly won't do here where it is envisaged that a blend is homogeneous. The elements cannot be defined as Aristotle defined them in terms of contrary extremes of warmth and humidity if it is possible for them to cooccupy a given region in view of Alexander's principle that contrary features cannot be exhibited in the same place at the same time. The central idea of the Stoic view is saved in modern chemistry by following the course provided for by thermodynamics: abandoning the Aristotelian homogeneity criterion of sameness of substance and adopting alternative criteria of a quantity's comprising a single substance, which are discussed in the next chapter.

5.7 Potential Parts and Potential Qualities

It might be said that the present treatment embodies a doctrine of actual parts which is at odds with a fundamental tenet of Aristotle's thought, namely the doctrine of potential parts. In Holden's words, "it is, in fact, a particular application of his famous view that the parts of continua are posterior to the whole and require some positive operation to actualize" (Holden 2004, p. 95). A viable account of such a doctrine would pose a challenge not only to the present interpretation of Aristotelian views and their rival Stoic conceptions, but to the way mereological ideas are developed throughout this book.

A question immediately arises of what is meant by "operation". Is it an operation in the mathematical sense in which the term is used here, or a process of removing parts? Ambiguities are compounded when Holden describes "the doctrine of potential parts" as the theory

> according to which the parts of bodies are merely possible or potential existents until they are generated by division. They represent ways in which the whole could be broken down, but do not exist other than as aspects of the whole until a positive act of division actualizes them as so many independent entities. Here the whole is a metaphysically simple (noncomposite) structure best described with mass terms: prior to division, it is not so many distinct things, but rather just *so much* metaphysically undifferentiated material (Holden 2004, pp. 79–80).

Clearly, additional concepts over and above strictly mereological and modal notions are involved here. The central idea of division would seem to be that of the process of removing a part from what remains of a body. An advocate of the potential parts doctrine might reject an analysis of removing a part of a body in terms of such notions as occupying, abutment and being distant from along the lines suggested in Sect. 2.3 above. But determining whether there is any genuine disagreement between the positive aspects of the doctrine of potential parts (as opposed to the blatant denial of actual parts) and the understanding of the mereology of bodies advocated here would clearly require an account of the associated notions.

The potential parts doctrine immediately faces the intuitive difficulty that it is strange to talk of parts after their removal. If a piece of matter is not part of another before being broken off, it is distinctly odd to say that it has become a part afterwards. A piece that is broken off is not something created by removal, but something already there which was susceptible to detachment. What would the removed pieces be parts of? If the whole is the scattered body resulting from division, why should the removed part be thought to be created by the division rather than the division being thought of merely as putting the parts at a spatial distance from one another? Why, in other words, should the scattered object be thought to be different, mereologically speaking, from the initially connected body? A similar issue arises with the suggestion that prior to division, the whole is "best described with mass terms ... [as] metaphysically undifferentiated material". Haven't we seen that Aristotle himself described undifferentiated material in *DG* I.10 in terms of a distributive condition, as being like a compound "any part of [which] ... is the same as the whole, just as any part of water is water"? The proponent of the potential parts doctrine cannot rest content with merely denying the actual existence of parts, without which there would be no distributive condition giving a nontrivial account of what it is for matter to be "undifferentiated" (whether metaphysically so, whatever that means, or not). We are owed an account of what is "best described with mass terms ... [as] metaphysically undifferentiated material" if it doesn't have a mereological structure of parts.

The potential parts doctrine has been resisted by posing a challenge to articulate a viable alternative rather than claiming to establish the actual parts doctrine on the basis of additional or "deeper" premises. In his review of the classical enlightenment arguments purporting to demonstrate the actual parts doctrine, Holden (2004, pp. 103–18) shows that they are all petitios, which wouldn't convert the advocate of the potential parts doctrine because the crucial assumption is exactly what they deny. No claim is made here to demonstrate the mereological principles appealed to at various places in the book on the basis of "deeper" premises. But the question arises of whether what proponents of the potential parts doctrine seek to deny by denying the actual parts doctrine is what is asserted here. One of the arguments Holden examines, forcefully pressed by Pierre Bayle (1647–1706), seeks to demonstrate the existence of parts on the basis of the principle that nothing can have incompatible qualities at the same time. Holden maintains this involves an equivocation, sliding from a sense of objects as subjects (which cannot sustain incompatible predicates), to which the premise applies, to a conclusion about objects "in the sense of independent beings" (p. 113). The "more felicitous version" of the principle about incompatibility, from which Bayle's version is ultimately derived, is found in Plato's *Republic*, and runs "the same thing will not be willing to do or undergo opposites *in the same part of itself*, in relation to the same thing, at the same time" (Holden 2004, p. 114; his emphasis). In this form, the principle only rules out

> the possibility that 'one thing' could have incompatible qualities at the same time, in the same respect, and *in the same part of itself*. And this principle is clearly preferable to Bayle's ... since it does not assume that if one thing supports incompatible qualities in different parts, then those parts are necessarily ontologically independent of one another (loc. cit.)

If this is what the actual parts doctrine is supposed to involve, then it goes beyond anything claimed here. What is claimed here is that macroscopic bodies have a mereological structure of parts in just the sense that is expressed in the principle of Plato just quoted. Like Aristotle's notion of undifferentiated material, Plato's principle makes explicit reference to the parts of a body. Mereological structure is not understood here to saddle its existential claims about the existence of parts with conditions about not requiring anything else in order to exist. In the sense of necessity of interest here, we have seen that quantities satisfy the Barcan principle and its converse. There is no possibility of quantities, and so their parts, possibly existing if they don't or possibly not existing if they do. If the doctrine of actual parts requires the parts to satisfy some such principle as continuing to exist when other parts might, in some sense, not, then the existential claims of mereological structure appealed to here do not involve the doctrine of actual parts. Perhaps it might be better simply to speak of mereology and avoid reference to the doctrine of actual parts. At all events, it is beginning to look as though not even the negative thesis of the potential parts doctrine denies the mereological structure of bodies as understood here, and doesn't form the basis of a genuine alternative to that view.

Holden (2004, pp. 118–27) goes on to discuss arguments purporting to demonstrate the potential parts doctrine and concludes that these are equally petitios, begging the question against the proponent of the actual parts doctrine. The most revealing of these arguments claims that according to the ordinary understanding of the notion, division creates several things from one, actualising parts rather than merely unveiling pre-existing parts, and the proponent of the actual parts doctrine misunderstands the concept of divisibility. There is, of course, something in this point, even if it doesn't make the case for the potential parts doctrine. It has been repeatedly emphasised here that it is a mistake to read too much into mereological concepts, and richer notions should be adequately developed on the basis of additional vocabulary and associated principles. Division is a case in point. In a particular context, it may be clear that a weaker notion abstracted from ordinary usage is at issue (as, for example, when axiom MT10 was described in Sect. 1.3 as expressing the infinite divisibility of time). We might well expect that the actual parts theorist should be understood in this way when speaking of the divisibility of space and time. But there is no getting away from the fact that the potential parts theorist is correct to point out that the ordinary notion of division is a richer notion that has application when speaking of material objects, and the mereologist should be aware that describing matter as infinitely divisible is ambiguous. This is not to grant the substantial claim of the doctrine of potential parts since the argument faces the corresponding conceptual difficulty noted above that it is strange to talk of parts after their removal and division can in any case be explicated as the removal of a part to a distant region on the basis of the standard mereological understanding of parts. There is no implication that the removed part is thereby created, although the creation of a new individual might be involved.

Can Aristotle's notion of the merely potential presence of elements in a mixt be dealt with analogously without calling on the doctrine of potential parts? Taking an anachronistic example involving just two elements, consider how water might be understood as a compound in Aristotelian fashion as derived from the combination of the elements hydrogen and oxygen. A first shot at expressing the idea that a quantity of water is potentially hydrogen and oxygen might run:

$$Water(\pi, t) \supset \Diamond \exists \rho \exists t' (\rho \subset \pi \wedge H(\rho, t') \wedge O(\pi - \rho, t')), \tag{5.11}$$

where H and O stand for hydrogen and oxygen, respectively. But in view of the Barcan formula and mereological essentialism, this is equivalent to

$$Water(\pi, t) \supset \exists \rho \exists t' (\rho \subset \pi \wedge \Diamond (H(\rho, t') \wedge O(\pi - \rho, t'))), \tag{5.12}$$

to the effect that there is a definite part of the water, π, that is possibly hydrogen and the remainder is possibly oxygen. Although this may well be a plausible rendering of the Stoic interpretation of water considered (anachronistically) as a blend and even the modern view, it may seem at odds with the Aristotelian idea of homogeneity. Why, on that view, should there be any particular part which is possibly hydrogen? Why, in other words, should there be a particular mereological partition into what is possibly hydrogen and oxygen? There would be no features in a homogeneous mixt distinguishing one proper part as possibly hydrogen and the remainder as possibly oxygen. Aristotle seems to have had some such thought in mind when he argued against the idea of a compound held by those who "conceive it as composition—just as a wall comes-to-be out of bricks and stones; and this mixture will be composed of the elements, these being preserved in it unaltered but with their small particles juxtaposed each to each". He continues

> That will be the manner, presumably, in which flesh and every other compound results from the elements. Consequently, it follows that Fire and Water do not come-to-be out of any and every part of flesh. For instance, although a sphere might come-to-be out of this part of a lump of wax and a pyramid out of some other part, it was nevertheless possible for either figure to have come-to-be out of either part indifferently: that is the manner of coming-to-be when both come-to-be out of any and every part of flesh. (*DG* II.7, 334a29ff.)

Just as a sphere and a pyramid can be obtained from *any* way of dividing a lump of wax into two parts of given relative size, so it must be possible to derive the elements from "any and every part of flesh". The Aristotelian conception of the potential presence of elements in compounds is one in which a given division into elements is possible, but doesn't exhaust the possibilities.

The idea of potential parts is not the only idea that stands opposed to the idea that definite parts are possibly this or that. There is also the idea that *any* part (of a certain kind) might fit the bill. The force of "any" is usually captured by a universal quantifier outside the scope of an appropriate operator. ("Any perception could be deceptive" is true iff every perception is such that it is possible that it is deceptive, in contrast to "every perception could be deceptive", which is true iff it is possible that every perception is such that it is deceptive.) But perhaps not literally any two-part partition into possible hydrogen and oxygen is possible. Immediately preceding the passage from *DG* I.10 quoted above where Aristotle says that a compound must be uniform, he also requires that "the part exhibit the same ratio between its constituents as the whole" (328a8–9). So a given partition is possible in which the elements exhibit a certain ratio, and any other division preserving this ratio of the elements is equally possible:

$$\begin{aligned} & Water(\pi, t) \supset \exists \rho \exists t' (\rho \subset \pi \wedge \Diamond (H(\rho, t') \wedge O(\pi - \rho, t')) \wedge \\ & \forall \sigma (\sigma \subset \pi \ \wedge \ SameRatio(\rho, \pi - \rho, \sigma, \pi - \sigma) \,.\!\supset \\ & \Diamond \ (H(\sigma, t') \ \wedge \ O(\pi - \sigma, t'))). \end{aligned} \tag{5.13}$$

"$SameRatio(\pi, \rho, \sigma, \tau)$" here abbreviates the relation of the ratio of the amount of matter π to that of ρ being the same as that of σ to τ. Note that mereological essentialism for quantities allows the "same substance" relation to express a rigid comparison between quantities which would not be possible otherwise. As for the criterion of sameness of size, the interpretation of Aristotle's few and unspecific references to material proportions is something of a mystery, bearing in mind that he had no notion of mass on the basis of which the law of constant proportions was formulated at the beginning of the nineteenth century. But without some restriction on the universal quantifier akin to that expressed by the "same ratio" clause here, the Aristotelian view of potential presence would be distinctly less attractive.

The Stoic view would modally qualify (5.12) in a different way, asserting that water is partly possibly hydrogen and partly possibly oxygen with a necessarily unique partition:

$$Water(\pi, t) \supset \exists \rho \exists t' (\rho \subset \pi \wedge \Diamond (H(\rho, t') \wedge O(\pi - \rho, t')) \wedge \\ \Box \forall \sigma (\sigma \subset \pi \wedge \Diamond (H(\sigma, t') \wedge O(\pi - \sigma, t')) . \supset \rho = \sigma)) \tag{5.14}$$

where again, mereological essentialism plays a role in ensuring the rigid identity. In theory, the Stoic view takes the possible hydrogen in water to be the same elemental substance as hydrogen actually occurring in water, and similarly for the oxygen, which might suggest simplifying (5.14) to

$$Water(\pi, t) \supset \exists \rho \exists t' (\rho \subset \pi \wedge H(\rho, t') \wedge O(\pi - \rho, t') \wedge \\ \Box \forall \sigma (\sigma \subset \pi \wedge H(\sigma, t') \wedge O(\pi - \sigma, t') . \supset \rho = \sigma)). \tag{5.15}$$

But it remains, as we saw, a mystery how elements can be generally characterised on the Stoic view as the same substance when combined as they are in the isolated state. (5.14) allows for a more cautious reading in which the elements are characterised by features exhibited in isolation, as Aristotle understood the characterisation of elements.

Chapter 6
The Nature of Matter

6.1 The Scope of Metaphysics in the Enlightenment

Aprioristic metaphysical thinking continued and continues to temper how philosophers think about the nature of matter after recognition of the contributions of empirical study in the Enlightenment. This is well illustrated by what Holden calls the argument from composition, which he discusses at length (Holden 2004, Ch. 4). The argument, purporting to show that matter comprises a ground floor of ultimate indivisible parts or atoms, was widely endorsed in the early modern period by rationalist and empiricist schools alike. It runs as follows

> Given the actual parts analysis of matter, material bodies are essentially composite or compound structures. We then add the claim that such complex, composite entities are ontologically derivative, depending for their existence on the prior existence of their parts. And if the parts are themselves also composite, then they in turn depend for their existence on the prior existence of *their* parts. But (according to the current argument) if the whole original is to exist at all, this ontological regress cannot go on forever with no ground floor. Given that the original composite whole exists, there must be an atomic base of noncomposite first parts whose own existence is not so derivative. (Holden 2004, pp. 170–1)

It is the pre-critical Kant's version of this argument that Holden finds the clearest and most telling. Kant's thesis, which he takes to be self-evident, is that a body is just a collection of distinct parts standing in a relation of composition which can be considered to be withdrawn or cancelled without detriment to the existence of the parts standing in that relation, leaving us with simple, uncompounded, bodies. "So bodies must ultimately be concatenations of these 'absolutely simple fundamental parts': metaphysical atoms" (Holden 2004, p. 174).

Holden discusses and dismisses several traditional objections to Kant's argument, which charge that it is a petitio, insisting that most of them evaporate once it is realised that the central tenet is the contention that the instantiation of a relation depends on the existence of the relata. This is why he could give the sense of the composition relation by describing it as concatenation. The physical import of

P. Needham, *Macroscopic Metaphysics*, Synthese Library 390,
https://doi.org/10.1007/978-3-319-70999-4_6

the composition relation holding the body together is irrelevant. All that matters is that it is a contingent relation, which Kant can envisage as not holding. Holden concedes that there is some force to the idea; "it is unsettling to think of this sort of ontological dependence running on forever with no *ultimate* relata, relata that do not themselves depend on further lower-level relations obtaining" (p. 184). But he grants that rejecting it "is a logically respectable position, although it will induce an unsettling metaphysical vertigo in many" (p. 205).

Another venerable tradition in philosophy has shown more interest in the nature of the relation holding matter together. Freudenthal (1995), for example, sees the quest for understanding the cohesion of matter to have been the fundamental question pursued in ancient theories of matter, and this seems to have been the more fruitful line of thought. Aristotle dismissed atomism because he took unchanging ultimate particles as conceived by the atomists to be incapable of interacting and we can certainly agree with this view without incurring metaphysical vertigo. A proponent of the infinite divisibility of matter (in the mereological sense, not raising questions of removing parts) might adopt this attitude, as might the agnostic who allows that further evidence could turn the scales either way. Attempting to circumvent considerations about the actual nature of atoms and introduce them by purely abstract, a priori reasoning raises questions about how such things could be incorporated into a workable physical theory dealing with the properties of matter. Like purely mathematical arguments, principles for which a priori argument provides the only motivation are not informative about the physical nature of matter. The Kant-Boscovich force-shell model of indivisible atoms as unextended points at the centre of a force field satisfies the principle. But it is not itself free from the metaphysical problems of the sort it was primarily devised to address, and although it might have been suggestive, it didn't itself amount to the basis of a viable physical theory of matter. The subsequent success of field theories was not connected with an atomic conception of matter.

Whilst philosophers debated the metaphysical status of atoms, others despaired of making any headway in understanding the nature of matter by entering into such considerations:

> if, by the term elements, we mean to express those simple and indivisible atoms of which matter is composed, it is extremely probable we know nothing at all about them; but if we apply the term elements, or principles of bodies, to express our idea of the last point which analysis is capable of reaching, we must admit, as elements, all the substances into which we are capable, by any means, to reduce bodies by decomposition. (Lavoisier 1965, p. xxiv)

By "decomposition" Lavoisier didn't mean a purely mental process of envisaging a division into simpler substances, but the actual performance of a physical process. Moreover, it was a search for elemental *kinds of substance*, not atomic individuals. Theorising on the basis of empirically sound reasoning seems to have furnished greater insight into the nature of macroscopic physical ontology of material bodies and processes than resorting to speculative metaphysical principles. Theorising of any kind involves conceptual structure with distinctive features, but these are revisable rather than fixed once and for all by a priori constraints, and leave questions open where the weight of evidence is inadequate or reason inconclusive. The development of a general view of a macroscopic physical ontology is pursued here in this spirit, guided more by general concepts and distinctions motivated by empirical rather than a priori considerations.

6.2 Phase-Bound Substances

Modern physics got off the ground in the seventeenth century with Newtonian mechanics, where matter is characterised with the universal property of mass. Aristotelian physics related the way different bodies move to their different chemical constitution, but Galileo and then Newton removed the distinctions of substance from the mechanics of motion. There was therefore no role for chemistry, concerned as it was with the distinction of kinds of substance, in the physics of the scientific revolution. But after Lavoisier incorporated the notion of mass at the end of the eighteenth century, chemistry finally became properly integrated with physics in the second half of the nineteenth century with the development of thermodynamics, which recognises a division of the mass into amounts of different substances. By contrast with the Aristotelian conception of substance, thermodynamics treats substance and phase independently, and it is interesting to note some points in the development of the notion of chemical substance after the advent of Newtonian mechanics leading up to the establishment of this distinction.

In 1761 Joseph Black discovered that heating a body doesn't always raise its temperature. In particular, he noticed that heating ice at 0 °C converts it to liquid at the same temperature. Similarly, there is a *latent heat* of vaporisation which must be supplied for the conversion of the liquid into steam at the boiling point without raising the temperature. Together with Black's earlier account of specific heat, this showed that heat must be distinguished from the state of warmth of a body and even from the changes in that state. But it was some time before the modern interpretation of Black's ground-breaking discovery was fully developed. Theories of heat as a substance in its own right were common at the time and Black was familiar with them, although it was Lavoisier who was to coin the term "caloric" for the heat substance. Black seems to have thought that what we would call the phase changes when ice melts and liquid water boils involve a combination of one substance with the heat substance. Thus, describing an experiment in which water is boiled, he says

> The water that remained could not be hotter than the boiling-point, nor could the vessel be hotter, otherwise it would have heated the water, and converted it into vapour. The heat, therefore, did not escape along with the vapour [on opening a valve], but in it, probably united to every particle, as one of the ingredients of its vaporous constitution. And as ice, united with a certain quantity of heat, is water, so water, united with another quantity of heat, is steam or vapour. (Quoted by McKie and Heathcote 1935, pp. 23–4)

Black evidently thought that supplying latent heat involved chemical reactions which, using Lavoisier's term "caloric", might be represented by

$$\text{ice} + \text{caloric} \rightarrow \text{water},$$
$$\text{water} + \text{caloric} \rightarrow \text{steam}.$$

Only free caloric, uncombined with any other substance, leads to the increase in a body's degree of warmth as it accumulates. This shows that Black was still in the grip of the Aristotelian conception of substance as necessarily connected with a certain phase: to change the phase is to change the substance.

Lavoisier took over this understanding of substances lock, stock and barrel, his reprimanding Aristotle's doctrine of the four elements notwithstanding (Lavoisier 1965, p. xxiii). He even retains one of the Aristotelian elements, listing caloric as the "element of heat or fire" (Lavoisier 1965, p. 175). This element "becomes fixed in bodies ... [and] acts upon them with a repulsive force, from which, or from its accumulation in bodies to a greater or lesser degree, the transformation of solids into fluids, and of fluids to aeriform elasticity, is entirely owing" (1965, p. 183). He goes on to define gas as "this aeriform state of bodies produced by a sufficient accumulation of caloric". Under the list of binary compounds formed with hydrogen, caloric is said to yield hydrogen gas (1965, p. 198), and he says that hydrogen is the base of hydrogen gas. Similarly, under the list of binary compounds formed with oxygen, caloric yields oxygen gas (1965, p. 190), and with phosphorus yields phosphorus gas (1965, p. 204).

In other respects, Lavoisier's list of elements departs radically from the Aristotelian four. Lavoisier criticised his predecessors for adhering to Aristotelian doctrine rather than leaving the number and nature of elements open to empirical investigation.

> the fondness for reducing all the bodies in nature to three or four elements, proceeds from a prejudice which has descended to us from the Greek Philosophers. ... It is very remarkable, that, notwithstanding of the number of philosophical chemists who have supported the doctrine of the four elements, there is not one who has not been led by the evidence of the facts to admit a greater number of elements into their theory (1965, pp. xxii–iii).

The criterion of an element as the last result of analysis (to which Aristotle also subscribed; see the passage from *De Caelo* III.3, 302a15ff. quoted in Sect. 5.4) leaves it open to empirical analysis to determine which substances are elements, and precludes the determination of the number and nature the elements by any independent route. Nevertheless, fire, in the form of caloric, remained an element in Lavoisier's scheme, but air is analysed into components. During the calcination of mercury, "air is decomposed, and the base of its respirable part is fixed and combined with the mercury ... But ... [a]s the calcination lasts during several days, the disengagement of caloric and light ... [is] not ... perceptible" (1965, p. 38). The base of the respirable part is called oxygen, that of the remainder azote or nitrogen (1965, pp. 51–3). Thus, oxygen is the base of oxygen gas, and is what combines with caloric to form the compound which is the gas.

Water was famously demonstrated to be a compound of hydrogen and oxygen by an experiment of Cavendish's, although first published by Watt, in which it is produced by combining hydrogen with oxygen by combustion. Interpretations of the experiment varied, Cavendish describing it as the burning of inflammable air in dephlogisticated air and concluding that dephlogisticated air was dephlogisticated water whilst hydrogen was either pure phlogiston or phlogisticated water. Lavoisier opposed phlogiston theories on the basis of a series of experiments, involving both analysis and synthesis, designed to justify his interpretation and rule out any others (1965, pp. 83–96). The last of these is mentioned here, illustrating the crucial principle involved in this interpretation of the experimental facts:

> When 16 ounces of alcohol are burnt in an apparatus properly adapted for collecting all the water disengaged during the combustion, we obtain from 17 to 18 ounces of water. As no substance can furnish a product larger than its original bulk, it follows, that something else has united with the alcohol during its combustion; and I have already shown that this must be oxygen, or the base of air. Thus alcohol contains hydrogen, which is one of the elements of water; and the atmospheric air contains oxygen, which is the other element necessary to the composition of water. (1965, p. 96)

The metaphysical principle of the conservation of matter—that matter can be neither created nor destroyed in chemical processes—called upon here is at least as old as Aristotle. What the present passage illustrates is the employment of a criterion of conservation—the preservation of mass. The total mass of the products must come from the mass of the reactants, and if this is not to be found in the easily visible ones, then there must be other, less readily visible reactants. This certainly wasn't the operational criterion it is often billed as being. Although the imponderable caloric doesn't conflict with the criterion, it is certainly not identified by virtue of the criterion. Light also figures in his list of elements, being said "to have a great affinity with oxygen, ... and contributes along with caloric to change it into the state of gas" (1965, p. 185). Like caloric, it too is without weight, but Lavoisier saw no obstacle in the criterion of decomposition to the introduction of elements which had a role to play in his scheme.

Not only was the postulation of caloric not justified by the principle of the preservation of mass. Lavoisier was also sensitive to a certain critique of the role of caloric in his theory. The elastic property of caloric, arising from its resistance to compression by the mutual repulsion of its parts, was supposed to explain the "aeriform elasticity" of gases, i.e. the tendency to expand and increase in pressure with temperature. But as he remarks,

> It is by no means difficult to perceive that this elasticity depends upon that of caloric, which seems to be the most eminently elastic body in nature. Nothing is more readily conceived, than that one body should become elastic by entering into combination with another body possessed of that quality. We must allow that this is only an explanation of elasticity, by an assumption of elasticity, and that we thus only remove the difficulty one step farther, and that the nature of elasticity, and the reason for caloric being elastic, remains still unexplained. (Lavoisier 1965, p. 22)

Duhem (1893a, p. 125) observed that the idea of a force is only delayed by the introduction of a force-carrying medium. But this lesson hadn't sunk in a century or so earlier when Lavoisier's contemporaries were grappling with the mysteries of matter.

6.3 Distinguishing Substance and Phase

According to Lavoisier's analysis, then, water is a compound of two elements, oxygen and hydrogen, which are the base of oxygen gas and base of hydrogen gas, together with a third element, caloric, the amount of which varies between solid, liquid and gaseous water. None of these elements did he claim to have isolated. Oxygen gas and hydrogen gas, on the other hand, are not components of water. This is very close to how the composition of water is understood today. When combined in a compound, the elemental components cannot be described as possessing a particular phase property of being solid, liquid or gas. Clearly, water in the liquid state doesn't consist of anything which is gaseous. Aristotle accommodated the point with his doctrine that the elements are potentially but not actually present in a compound. It is unclear what the Stoic view of the elements was, but we saw that their view of blends requires that the component elements don't retain all the properties they exhibit in isolation, and this would include the phase properties. Although Black still adhered to the Aristotelian conception of substances as essentially possessing one specific phase property, Lavoisier took a step away from this with his notion of an elemental component as something which is common to the solid, liquid and gaseous states of the element. But because he maintained Black's caloric theory of latent heat, the elements were substances not occurring in isolation; they are bound at least with more or less caloric in one or other state of aggregation. A compound of elements, like water, forms a secondary compound with caloric in the liquid and gaseous states.

Lavoisier understood water to be composed of definite proportions of hydrogen and oxygen as a result of his analyses conducted in the 1770s. What was to become formally named as the law of definite proportions in the first decade of the nineteenth century, after Proust was generally acknowledged to have won a protracted dispute with Berthollet, was thus already widely accepted by leading chemists. Perhaps it was Berthollet's championing the contrary view, that the strength of chemical affinity between given elements varies according to circumstances, that forced chemists to explicitly acknowledge and defend the principle of constant proportions. At all events, the law was taught as part of chemical theory from the early nineteenth century, and provided a criterion on the basis of which a definition of compounds could be formulated. It decided the issue between Davy, who thought air to be a compound, and Dalton, who took it to be a homogeneous mixture with variable composition. It was supposed that whilst the component elements of compounds were chemically combined, the component substances of homogeneous mixtures (called solutions) were merely mechanically mixed. In both cases, the properties of the original components are changed—the compound sodium chloride is neither a soft grey metal like sodium nor a greenish-yellow gas like chlorine (at room temperature and pressure), and brine is neither a white solid like sodium chloride nor a tasteless liquid like water. But the law of constant proportions only gave a criterion of chemical combination; it provided no explanation of what chemical combination amounts to.

When the law was first formulated, the caloric theory of heat still reigned. But by mid century the basic tenet of the theory, that caloric is conserved in all processes, was disproved with the realisation, embodied in the first law of thermodynamics, that heat is not preserved but is interconvertible into mechanical work. Ontologically, the notion of caloric as a substance gave way to the existence of processes—in particular, heating, which changes the state of warmth of a body—and the new concept of energy, in terms of which an equivalence between heating and mechanical working is determined as a measure of the amount of heating in principle obtainable from a given amount of working. The Black-Lavoisier account of phase change was no longer tenable and the distinction between, for example, base of oxygen and oxygen was undermined. But Lavoisier's insight that a compound like water can't be said to be composed of elements associated with particular phases still held good: water is solid or liquid under conditions that hydrogen and oxygen are gases. The statement of the law of definite proportions requires a concept of substance which is independent of phase, and a substance, be it an element or a compound, is not in general bound to a particular phase. Phase changes requiring input of a certain latent heat are changes in one and the same substance from one phase to another. (Some compound substances are not stable under phase change; ammonium chloride, for example, decomposes on heating under normal pressure rather than changing from solid to liquid or gaseous ammonium chloride, and biologically important substances such as proteins typically decompose before reaching a melting or boiling point.)

The first law of thermodynamics places no restriction on the equivalent interconversion of states by heating and working beyond the principle that the amount of energy, in terms of which each is measured, is conserved. But many states which are possible as far as the first law alone is concerned are precluded by a further restriction imposed by the second law of thermodynamics, which forces change. Just as a kettle of hot water cannot remain hot indefinitely in a room at ordinary temperature and pressure, so chemical substances such as metallic sodium and water cannot remain as such when in contact. The hot water in the kettle cools down to room temperature, and the sodium and water undergo a chemical reaction generating sodium hydroxide and hydrogen. But these processes are not required by the first law. It is the second law, which provides the foundations for the concept of entropy, that lays down the principle driving these processes. Developments in thermodynamics culminating in Gibbs' groundbreaking "On the Equilibrium of Heterogeneous Substances" (1876–1878) provided a theory of stability which was the first workable general theory of chemical combination and remains a bulwark of modern chemical theory.

Daltonian atomism promises an understanding of compounds which is more easily grasped than the macroscopic theory of thermodynamics. But Dalton's conception of atoms of the same elemental kind which repelled one another in virtue of a surrounding atmosphere of caloric became an embarrassment which chemists soon gave up. Daltonian atomism came to be understood as an abstraction from what Dalton actually said about the nature of atoms, which its nineteenth century critics chided for not providing a positive account of what atoms were like which would indicate how they could combine to form molecules and crystalline structures. Not until 1927 was there the remotest suggestion of how atoms can actually combine. Lewis (1916) had criticised Bohr's (1913) symmetric atom as offering no foundation for molecules with atomic bonds oriented in various direction is space, and suggested that bonding is a matter of electron pair sharing (covalent bonding) or electron transfer in the formation of ions. But how like-charged electrons could hold atoms together in chemical combination rather than making them repel one another (as did Dalton's atoms in virtue of their atmospheres of caloric) was not answered before the application of quantum mechanics to the problem by Heitler and London (1927). Molecules were a nineteenth century mystery. They are completely impossible entities according to nineteenth-century physics. Whether the mystery has been finally laid to rest with the advent of quantum mechanics is still disputed (see Woolley 1988; Sutcliffe 1993). But it belittles the distinctive achievement of quantum mechanics to suggest that there was a viable atomic theory of chemical combination, or even a plausible hint of how atoms could combine, before 1927.

6.4 Thermodynamics and the Phase Rule

It is remarkable that a theory devised for the purpose of finding the theoretical principles on which steam engines operate should have a bearing on substances and their transformations into other substances, let alone play a central role in the determination of these concepts. But that is precisely what happened with the developments in thermodynamics culminating in Gibbs' work noted above. Like Aristotle, Gibbs allowed that a quantity of matter might comprise different homogeneous bodies, formed from different component substances, constituting a heterogeneous mixture. But whereas Aristotle thought that each homogeneous body in a heterogeneous mixture comprises a different substance, Gibbs allows that they may themselves be mixtures composed of some or all of the substances present. Gibbs therefore needed a basic distinction between substances and the different homogeneous bodies constituting a heterogeneous mixture. He calls the different homogenous bodies which "differ in composition or state[,] different phases of the matter considered" (Gibbs 1948, p. 96). In general, a mixture comprises a number coexisting phases over which several substances are distributed, and when at equilibrium it is governed by a general law which has come to be known as Gibbs' phase rule. This states that

$$Variance = c - f + 2 \geq 0,$$

where c is the number of substances or components in the mixture, f is the number of phases, and the variance is the number of independently variable intensive magnitudes characteristic of the state of the whole system.

To illustrate, a 2-phase system comprising a single substance is, according to this law, univariant. Accordingly, the state of a system comprising water in the form of liquid in contact with the vapour enclosed in a container fitted with a piston allowing change of volume is governed by just one independent variable. So at a given temperature, the pressure is fixed. Any attempt to vary it (whilst the temperature is held fixed), for example by reducing the volume by depressing the piston, will only result in material being transferred from the vapour to the liquid, maintaining the pressure—as long as two phases remain. Similarly, attempting to reduce the pressure by withdrawing the piston and increasing the volume will only result in material being transferred from the liquid to the vapour, again maintaining the pressure. If the piston continues to be raised, assuming the container is sufficiently large, all the liquid eventually becomes gas, leaving only a single phase and the system becomes bivariant. Pressure and temperature can then vary independently, as allowed by the familiar gas law. This behaviour is *characteristic of a quantity of matter comprising a single substance*, and the specific data (the specific pressure as a function of temperature whilst there are two phases) is characteristic of the specific substance.

The phase rule allows another circumstance for a single-component system ($c = 1$), namely that the number of phases, f, is 3. Then the variance is 0, and heating will not alter the temperature or pressure until the solid disappears and there are just two phases. This is called a triple point, and for copresent solid, liquid and gas phases of water at equilibrium, the temperature is 0.01 °C and the pressure 4.58 mmHg (0.0060373 atm). (The triple point state is used to calibrate the absolute temperature scale, fixing a temperature above zero Kelvin which is assigned the value 273.16 K to bring the Kelvin scale into line with the Celsius scale.) This adds to the general behaviour characteristic of matter comprising a single substance, and again the specific data distinguishes the substance water from others.

Several factors bear on the interpretation of the phase rule, which can make its application not entirely straightforward. Some further insight into the significance of "c" in the Gibbs phase rule can first be got by noting how the law is derived. As the terminology of independent variables in the concept of variance suggests, the derivation of the phase rule is a matter of how many variables are determined by the available relationships expressible as equations. Gibbs reasoned that at equilibrium, the temperature and pressure have the same value at all points of the mixture (hence the "2"), as does the intensive property of each of the c components called its chemical potential. This determines the change in amounts of that substance, and is defined in thermodynamics analogously to the way temperature and pressure are defined.[1] But equality of chemical potential across the phases doesn't entail equality of concentration from phase to phase. These $c+2$ intensive variables cannot vary independently, however, because of a relation (which came to be called the

[1]In the so called energy representation, in which the internal energy is a primitive function, $U(S, V, N_1, \ldots, N_r)$, of the entropy, S, the volume, V, and the amount of the ith substance, $N_i, 1 \leq i \leq r$, the temperature is defined as $\partial U/\partial S$, the pressure is defined as $-\partial U/\partial V$ and the chemical potential of the ith substance, μ_i, is defined as $\partial U/\partial N_i$.

Gibbs-Duhem equation) holding for each of the f phases, providing connections eliminating f variables. Clearly, there cannot be a negative number of variables determining the state of a body, and the variance is always non-negative for a body in a condition that can be sustained at equilibrium.

Another derivation is often given in textbooks (e.g. Denbigh 1981, pp. 184–5) which doesn't explicitly call upon the Gibbs-Duhem equation. The starting point is that specifying the state of a mixture comprising several substances distributed over a number of phases requires the composition, i.e. the concentration, $x_{i\alpha}$, of each substance i in each phase α, to be given in addition to the temperature and pressure. These concentrations are expressed as proportions (mole fractions or weight percentages) that each substance contributes to the matter comprising each phase, so that $\Sigma_i x_{i\alpha} = 1$ for each phase α, whose composition is therefore specified by $c - 1$ independent variables. Together with temperature and pressure, then, each phase is specified by $c+1$ variables, given a total of $f(c+1)$ for the entire system. At equilibrium, however, a number of equalities hold between the variables. There are $f-1$ equalities between the temperature of each phase, and similarly $f-1$ equalities between the pressure of each phase. Furthermore, the chemical potential of the ith substance (itself a function of the temperature, pressure and $c - 1$ independent concentration variables for the phase) has the same value in each phase, so that for the c substances there are $c(f - 1)$ equalities $\mu_{i\alpha} = \mu_{i\beta} = \ldots$ across the f phases $\alpha, \beta, \ldots$. The total number of independent variables is therefore reduced by $(c + 2)(f - 1)$ equalities. Now if some of the independent equations connecting a set of variables are not to be incompatible, their number cannot exceed the number of these variables. Accordingly,

$$f(c + 1) \geq (c + 2)(f - 1),$$

or

$$c + 2 - f \geq 0.$$

There can therefore be at most two more phases than there are substances. A single substance can exhibit a maximum of three phases, at its triple point. The total number of independent variables is the difference between the total number of variables and the number of equations:

$$\textit{Variance} = f(c + 1) - (c + 2)(f - 1) = c + 2 - f.$$

The phase rule as stated presupposes that no further constraints are acting on the mixture which would further reduce the number of independent variables. But if there are r reactions taking place between the substances, then at equilibrium there will be an additional equation ($\Sigma \nu_i \mu_i = 0$, where ν_i is the stoichiometric coefficient—either $+a_i$ or $-b_i$—of the ith substance in the reaction

$$a_1 A_1 + \cdots + a_m A_m \rightleftharpoons b_{m+1} B_{m+1} + \cdots + b_n B_n$$

and μ_i is its chemical potential) for each of the r independent reactions taking place. These equations reduce the $c - f + 2$ variables by r, although this is normally put by saying that it reduces the number of independent substances, c, by r. Moreover, chemical reactions may take place in such a way as to render true further equations governing the composition of a phase, yielding additional restrictions. Where there are s stoichiometric constraints connecting the proportions of products of chemical reactions in a particular phase or several phases (these conditions may be intraphase relationships or interphase relationships), the variance = $n - f + 2 - r - s$. The number of independent substances, c, can be identified with $n - r - s$, but although this number is unique, which they are is usually not, and there are frequently several ways of selecting c independent substances from all those apparently present.

A good illustration is provided by heating the red oxide of mercury in a previously emptied closed container (Duhem 1910, pp. 175–6), which dissociates as follows:

$$2HgO \rightleftharpoons 2Hg + O_2.$$

Two phases appear, a solid comprising HgO and a gaseous mixture of oxygen and mercury vapour. There are three substances, but one equilibrium constraint and the further constraint that the gas mixture contains equivalent proportions of mercury and oxygen. So there is one independent substance, which means that the two-phase system is univariant: at 520 °C, for example, the pressure is fixed at 4.176 mmHg. Destroying the equivalent proportions by introducing either a little oxygen or a little mercury removes a constraint and there are two independent substances, making the system bivariant. Fixing the temperature no longer suffices to fix the pressure. Adding sufficient mercury to form a third, liquid, phase, however, returns the system to univariant behaviour. But now the pressure determined by a temperature of 520 °C is 8.440 mmHg.

Another example Duhem discusses is the classic case of the equilibrium

$$CaCO_3 \rightleftharpoons CaO + CO_2$$

established when calcium carbonate is introduced into a closed and otherwise empty container and heated. There are three substances and one equilibrium, and so just two independent substances. The system seemed to comprise two phases, a white solid and a colourless gas. But this couldn't be reconciled with its variance, which was determined to be one (there is a definite gas pressure for a given temperature). Compliance with the phase rule led chemists to believe there must be three phases—the gas phase, and two distinct solid phases, one comprising $CaCO_3$ and the other CaO. A single solid phase would be an intimately mixed solid solution of $CaCO_3$ and CaO, but they cannot form such an intimate mixture and there must be distinct calcium carbonate and calcium oxide phases. Subsequently, new methods of analysis of the crystal structure were able to show that the calcium oxide is confined to isolated "islands" within the surrounding calcium carbonate, much as oil forms droplets in water when a vinaigrette dressing is shaken.

Duhem (1898) pointed out that this is an example in which the assumption in the above derivations of the phase rule, that all the substances in the system are distributed over all the phases, doesn't hold. The gas phase contains neither calcium carbonate nor calcium oxide. But as he showed, a rigorous proof of the phase rule can be given which doesn't rely on this assumption.

The calcium carbonate system can be contrasted with that formed by mixing potassium sulphate, ammonium sulphate and insufficient water to dissolve them completely, which is trivariant. Not only can the temperature and pressure vary independently, but also the relative mass of one of the three substances. As might be expected, increasing the mass of one of the salts without changing either temperature or pressure changes the composition of the solution by increasing the concentration of the added salt and decreasing the concentration of the other salt. Since there are no chemical reactions the three substances are independent, and the number of phases, f, given by $3 = 3-f+2$, is 2, so that apart from the liquid solution there is only one solid phase, which, in contrast to the white solid in the previous case, is therefore a homogeneous solid solution of the two salts. In the case of the calcium carbonate system, the phase rule gives us that $1 = c - 3 + 2$, or $c = 2$. The terminology might suggest that, since the one solid phase comprises calcium carbonate and the other calcium oxide, we've necessarily used up our quota, leaving the gas without a substance composing it! But a distinction must be maintained between substance and phase. The calcium carbonate phase is not the same thing as the calcium carbonate substance. Further, thermodynamics doesn't select a specific pair from the candidates as the two independent substances. An alternative description, compatible with the phase rule, is that the two substances are CaO and CO_2, and the remaining substance is a combination of these two. Or again, $CaCO_3$ and CO_2 might be chosen, when the CaO must be understood, as Castellan (1964, p. 247) puts it, as $CaCO_3$ "minus" CO_2. This raises questions about what the various phase and substance predicates apply to, which are pursued in the next chapter.

A further example, studied by Zernike (1951, 1954), involves heating ammonium bicarbonate, NH_4HCO_3, in an initially otherwise empty container, when a dissociated vapour phase and a liquid phase appear in addition to solid ammonium bicarbonate. There is an interphase stoichiometric condition at play in this case, affecting the composition of several phases. The vapour phase contains ammonia, carbon dioxide and water, each of which have different solubilities in liquid ammonium bicarbonate. Consequently, neither the vapour phase (from which the substances dissolved in the liquid derive) nor the liquid has a composition in which the proportions are as represented by the formula NH_4HCO_3. The additional condition giving the proportions of the products of dissociation in the liquid and gas phases, which can be readily calculated, is counted as one of the conditions s that has to be taken into account in applying the phase rule.

Application of the phase rule is not, in general, a straightforward matter. These considerations suggest that even in apparently straightforward cases, it would be necessary to ensure that there are no complicating factors lurking behind the scenes. For example, enantiomorphs, i.e. isomers (different compounds with the same composition) whose molecules are mirror images of one another, are usually counted different substances. But they have the same values for their intensive properties except for rotatory power (equal in magnitude but opposite in sign), which may render some of the Gibbs-Duhem equations for different phases in a system no longer independent and therefore affect the variance (Scott 1977; Wheeler 1980). And apart from theoretical complications of this sort, experimental work always entails difficulties of its own. In this case,

> independently variable components are often discovered even in very thoroughly investigated substances, be it for the reason that single phases have been overlooked because of their small quantity [i.e. size] or of their similarity with others, or that the observed number of independent variations is too small. (Wald 1896, p. 26)

The dangers of systematic and experimental error are a fact of life for empirical investigation, and use of the phase rule is not exceptional.

To round off with a couple more examples, at high temperatures, water begins to dissociate. At around 2000 K (approximately 1700 °C) it is about 1% dissociated (Denbigh 1981, p. 153), and the equilibrium

$$2H_2O \rightleftharpoons 2H_2 + O_2$$

is established. This is a gas phase reaction in which the number of substances present apparently increases from one to three as the temperature rises. At the same time, the number of constraints on the whole system increases by two (the equilibrium and the equivalent proportions of H_2 and O_2). The system is bivariant, just like water in a single phase described above, and the number of substances in the sense of the phase rule remains one. A similar example is the dissociation of ammonia, where the following equilibrium is established at high temperatures

$$2NH_3 \rightleftharpoons N_2 + 3H_2$$

with the production of nitrogen and hydrogen. Again, what sense there is in saying that there is just the one substance, water, present despite the dissociation will become clearer in connection with the discussion of what the phase and substance predicates apply to in the next chapter. We will see that these two examples are formally like the dissociation of water into hydrogen and hydroxyl ions described in the next chapter, although they involve the macroscopically recognisable substances hydrogen and oxygen in addition to water in the one case, and ammonia, nitrogen and hydrogen in the other, rather than microscopic species. There we will see that the natural interpretation is that there is just one substance present, and the phase-rule notion of an independent substance is identified with the unqualified notion of substance. By parity of reasoning, there is a strong case for saying that the same interpretation should apply here, and just one substance be deemed to be present despite the dissociation.

Since these are equilibria, they could equally be established by starting with equivalent proportions of hydrogen and oxygen, or of nitrogen and hydrogen, as initial ingredients. This is then reminiscent of Aristotle's conception of mixing.

according to which when the original ingredients are mixed in "comparable amounts", a single substance results. As we saw, where "combination has taken place, the compound must be uniform—any part of such a compound is the same as the whole, just as any part of water is water" (*DG* I.10, 328a10f.). Homogeneous matter resulting from mixing would comprise a single substance. Despite the fact that Aristotle's general view of mixture clearly stands corrected by Gibbs' view, Duhem thought that thermodynamics showed how "chemistry [had] recover[ed] ... the peripatetic notion of a mixt" (1902, p. 183). The antiquated term "mixt" that Duhem uses for the Aristotelian conception of a uniform body resulting from a process of mixing avoids any suggestion that Aristotle recognised the distinction between a compound and a solution—a homogeneous (single-phase) quantity containing several substances. A solution is what we would have in our two cases at lower temperatures in the absence of a catalyst, when eons could pass before equilibrium is established after mixing equivalent proportions of hydrogen and oxygen, or of nitrogen and hydrogen. For appreciably shorter times of macroscopic order, the mixture can be treated as a homogeneous gas comprising two substances. Aristotle's notion of a mixt seems closer to the idea of a compound, for which it provides a simple model that serves as a foil illuminating more complex cases.

Elemental names are used in these cases with some ambiguity. Oxygen may occur as free oxygen or dioxygen, O_2, and as ozone, O_3. A mixture of the two sufficiently cooled and compressed that it becomes a heterogeneous, two-phase quantity is bivariant (Ricci 1951, p. 10). Unlike the univariance exhibited by a two-phase mixture of pure water, the phase rule indicates that two substances are to be distinguished, for which we need two names. Water is composed of hydrogen and oxygen, but not of the substances whose formulas are H_2 and O_2, appropriately named dihydrogen and dioxygen. And as we have just seen, the phase rule would suggest that dihydrogen and dioxygen are not substances occurring for ordinary macroscopic lengths of time in a quantity of water raised to 2000 K either. The situation is different for the low-temperature solution, which can be partitioned into dihydrogen molecules and dioxygen molecules.

These examples illustrate the interplay between substances and phases. They are cases where a quantity of matter exhibits the two kinds of properties, of a certain number of independent substance properties and a certain number of phase properties. In general, where there are several substances, the corresponding substance predicates are instantiated by proper parts of the entire quantity, but the same cannot be so straightforwardly said of the phase properties. A more careful statement calls for a more detailed discussion of phase and substance predicates.

Chapter 7
The Relation of Macroscopic Description to Microstructure

7.1 Water is H_2O

Matter changes with respect to the substance and phase properties that it bears. Water, as we have seen, is created and destroyed in various chemical processes. It is produced in what organic chemists call condensation reactions and in the combustion of hydrocarbons such as methane. It can even be directly generated from its constituent elements by burning, or mixing in the presence of a suitable catalyst. Water is destroyed when sodium reacts with it to produce sodium hydroxide and hydrogen at the expense of the corresponding (in accordance with the laws of chemical proportions) amount of water, and it is directly decomposed into its constitutive elements by electrolysis. A literal interpretation of creation and destruction of quantities of water might seem to accord with ordinary ways of speaking about chemical processes. But an interpretation more in line with Lavoisier's fundamental principle of the indestructibility of matter would naturally have it that quantities of matter may bear features of being such-and-such a compound substance for one time and not for another. Because processes proceed more or less gradually over a period of time, a quantity won't be all the initial substance after commencement of the transformation, and so won't be the initial substance. It will be partly the initial substance, in the sense that part of it will be the initial substance, and so the process continues until none of the quantity in question is the initial substance. Whereas substance properties typically hold temporarily, the quantities of matter that bear them exist permanently, sustaining the standard mereological relations and operations and making no call on time-dependent variants in the manner of Thomson (1983). Time-dependence is incorporated in the substance predicate.

Phase properties, the more common ones expressing the conditions of being solid, liquid or gas, are also time-dependent, and expressed by dyadic predicates applying to a quantity and a time. An ordinary phase change, occurring, for example, when an ice cube in a gin and tonic melts, might again seem to call upon the literal creation and destruction interpretation. But as with chemical transformations, the

P. Needham, *Macroscopic Metaphysics*, Synthese Library 390,
https://doi.org/10.1007/978-3-319-70999-4_7

quantity initially all ice is after some time only partly ice, the remainder being liquid. In this case, according to standard chemical theory, the quantity preserves the substance property; both ice and the liquid formed by melting are water. Some philosophers follow Putnam (1975, p. 225) in restricting being the same substance water to being the same liquid, but here we follow elementary scientific practice which teaches that *water* melts at 0 °C. In contrast to the phase-dependent sense of "water" implying that whatever it is true of is liquid, this is the sense of "water" in which H_2O is water. Looking more deeply into this connection provides some insight into how the quantity to which the "water" predicate applies for macroscopic times should be understood.

Since "water" is a dyadic predicate of a quantity and a time, so too is the "H_2O" predicate and "All water is H_2O" means that the two predicates apply to the same things at the same time:

$$\forall \pi \, \forall t \, (Water(\pi, t) \; \equiv \; H_2O(\pi, t)).$$

"Is H_2O" cannot apply to a single, isolated water molecule because "is water" applies to matter that freezes at 0 °C and boils at 100 °C, which it doesn't make sense to say of an isolated water molecule. Does it apply to some collection of appropriately related water molecules of sufficient size, leaving aside how "appropriately related" and "sufficient size" are to be spelt out so as to cover solid, liquid and gas phases and presence of other substances in solution? In the absence of a special context justifying the use of this short expression as a convenient abbreviation, this is a far-fetched suggestion, all the more so since there is a much more plausible interpretation. The expression "H_2O" is used as a compositional formula, which merely expresses the proportion of the constituent elements in a compound. In this sense, its significance is entirely macroscopic. The gravimetric ratio of hydrogen to oxygen of 1:8 is converted into a different scale, reflecting combining power, which can be understood as saying that there is twice as much hydrogen as there is oxygen in water. The analogous compositional formula CO_2 for carbon dioxide expresses on the same scale that there is twice as much oxygen in carbon dioxide as there is in water, despite the gravimetric carbon-to-oxygen ratio of 3:8, but just as much oxygen in water as there is carbon in carbon dioxide.

Compositional formulas were elaborated by the development of structural formulas in the course of the nineteenth century with the discovery of isomers, which are distinct substances with the same composition. Structural formulas accommodate several kinds of isomerism without making any direct commitment to microstructure,[1] and could be understood to represent the chemical properties of a substance, i.e. the general lines along which it will react with other substances. This serves to emphasise the point that in order to speak of microstructure, it is necessary to go beyond this and offer more specific descriptions of the microparticles constituting a substance if that is that we are interested in. These particles are

[1]As Duhem carefully explained in Duhem (1892, 1902); see Needham (1996, 2008a).

entities distinguished by internal bonds holding the parts together significantly more strongly than these parts adhere to adjacent material for an interval of time during which the particle exists.[2] The late nineteenth-century understanding of the molecular constitution of substances is still proffered as typical of many organic substances such as hydrocarbons. According to this conception, there is a single species of particle, the individual molecule, which the structural formula of the substance is held to represent, and the substance can be thought of as a collection of molecules. Although there may be intermolecular interactions, keeping the molecules at a distance and determining the overall macroscopic density, these modify the molecules rather than forming different kinds of particles from them. Further description would enter into the geometric features of the molecules, the bond strengths, and so forth. But as inorganic chemists in particular point out, what is typical for organic substances is not typical for substances in general. The principle of fixed valency governing the construction of structural formulas for organic substances does not hold in general. Moreover, if the microstructure is primarily a collection of particles, these particles are not always molecules in the sense of what a structural formula might be considered to represent, and may well not be all of the same kind or stable over macroscopic intervals of time. Water, in particular, is one of the substances which doesn't fit the organic substance paradigm at all.

7.2 What Do Substance Predicates Apply To?

Although the present concern is with macroscopic concepts, consideration of the microrealm has a bearing on what substance and phase predicates apply to. This is not simply a matter of a correlation of changes at the two levels: there is always change at the microlevel, which may involve more than translational, rotational and vibrational motion mentioned earlier, whether or not there is any change in macrofeatures. Although change at the macrolevel is matched by change at the microlevel, the same goes for stable equilibrium at the macrolevel, which doesn't correspond to a static state of affairs at the microlevel but rather continual fluctuation and change in which certain processes are countered by others in a dynamic equilibrium in which an overall balance obtains over sufficiently long times. The steady temperature of an ideal gas, for example, governs a given distribution of the energy over the gas molecules, so that a fixed number have energy within a given range as energy is continually exchanged between molecules. Even at the absolute zero of temperature, the indeterminacy principle of quantum mechanics requires a "zero-point" vibration. But as noted, the changes may well involve more than translational, rotational and vibrational motion.

[2]"A molecule may ... be defined as a small mass of matter the parts of which do not part company during the excursions which the molecule makes when the body to which it belongs is hot" (Maxwell 2001, p. 305).

Since substance and phase predicates are dyadic, applying to a quantity at a time, the question arises of what the nature of the quantity at issue is for the time in question considered in the light of the underlying dynamic situation at the microlevel. But macro- and microlevel descriptions differ in general terms not only in the contrast between an apparently static equilibrium or a stately rate of change at the macroscopic level and a considerably more dynamic microlevel. Microentities are not simply smaller versions of macroentities following essentially the same laws with essentially the same kind of features such as occupying definite regions for definite intervals of time. The quantum realm is populated by entities of a radically different kind, subject to ever greater indeterminacy in energy for ever shorter times when not in a stationary state. The import of the distributivity condition introduced in Chap. 4 as proper parts are pursued ever closer to the quantum domain is anything but straightforward. The discussion in Chap. 4 led to the suggestion that the variables to which substance, phase and related predicates apply range over a domain of macroscopic entities. Two lines of defence of the distributivity condition for substance predicates will be explored here. In a later section, the typically quantum mechanical feature of indistinguishability is used to restrict the parts of macroscopic quantities so as not to include entities of the quantum domain. But first, the import of the dynamic situation at the microlevel for what substance and phase predicates apply to is discussed without considering the quantum character of microentities.

If microscopic considerations are to throw any light on the issue of what substance and phase predicates apply to at macroscopically significant intervals, it must be appreciated that it is not, in general, true that quantities of matter are just collections of molecules in the sense of mereological sums which can be partitioned into molecules (when the molecules would be mutually separate parts whose sum exhausts the quantity) that are preserved over macroscopic times. Substances which can be so partitioned are called *molecular*. Although there are substances, like methane, which are molecular substances, many substances are not molecular and water is a notorious example of one which isn't. Understanding how these mereological sums are taken over time provides some insight into how the continuous, macroscopic conception of matter ("gunk"?) is reconciled with the discrete microscopic conception.

Variation in microstructure is postulated in the first instance to explain variation in macroscopic features such as melting and boiling points and heat capacities. The molecular substance methane, with formula CH_4, is a typical organic hydrocarbon, comprising molecular species in which a carbon atom is strongly bonded to four hydrogen atoms. Weak, so-called van der Waals, cohesive forces act between the molecules and these forces are easily disrupted by thermal motions at relatively low temperatures. Such molecules form solids which are soft with low melting points that increase with molecular weight. The methane molecule is light and methane is not even a liquid at normal temperature and pressure, but a gas boiling at $-161.5\,°C$. Normal-pentane, a hydrocarbon of essentially the same kind (an alkane) with compositional formula C_5H_{12}, is liquid at room temperature boiling at $36.1\,°C$, and *n*-eicosane, with compositional formula $C_{20}H_{42}$, is solid at room temperature melting at 37 °C. These stand in contrast to substances like common salt (NaCl) and silicon dioxide (SiO_2), which form indefinitely repeating arrays in the solid phase held in place by considerably stronger forces than the van der Waals forces. In these substances, no particular pairs, in the case of NaCl, or triples, in the case of SiO_2, of atoms can be distinguished from their neighbours by virtue of stronger internal bonding between the atoms in putative molecules than between adjacent atoms. A lump of salt consists of an ionically bonded array in which each sodium cation (i.e. a positively charged ion) is surrounded by six chloride anions (i.e. negatively charged ions), and each chloride anion is surrounded by six sodium cations, in a structure which is repeated throughout the whole macroscopic crystal. No particular pairing of a sodium cation with one of its six coordinated chloride anions presents itself as more natural than a pairing with any of the remaining five anions.[3] The formula "NaCl" simply represents that there are as many equivalents of sodium as there are of chlorine in any sample of salt. Given the additional information about the microstructure that it is an ionic compound, the formula translates into microscopic terms as equal numbers of sodium and chlorine ions. Taking the formula to cover salt dissolved in the sea would entail the same equality of numbers of ions although there would be nothing corresponding to the microstructure of the solid crystal. Solid silicon dioxide also takes the form of an array extending over the entire macroscopic crystal in which each Si atom is bonded to four oxygen atoms (covalently, but with considerable ionic character) and each oxygen atom to two silicon atoms. Structures of these kinds are typically hard (solids at normal temperatures and pressures) with high melting points.

Water falls squarely between these two categories, with molecular units distinguished by internal bonding between a single oxygen atom and two hydrogens which is stronger than intermolecular attraction, but where the intermolecular attraction between two water molecules is midway in strength between a van der Waals interaction and an ionic bond. This explains water's unexpectedly high melting and boiling points compared with other substances of comparable molecular weight. Methane, for example, melts at −183 °C and boils, as we saw, at −161 °C, which gives some indication of the strength of the intermolecular force in water.

The intermolecular forces give rise to an indefinitely repeating crystalline structure in ice analogous to that described for common salt and silicon dioxide. Water has a surprisingly low latent heat of fusion (heat required to change the solid to liquid at the melting point) compared with the latent heat of evaporation (only 15% of the latter), suggesting that much of the intermolecular structure in the solid is in some sense retained in the liquid and not finally lost until in the gas phase. On the other hand, liquid water has a large heat capacity—nearly twice that of ice (at the melting point), and more than twice that of steam (at the boiling point)—

[3]Armstrong (1927, p. 478), then a retired professor of chemistry, became something of a laughing stock when he repudiated Bragg's claim that in sodium chloride there are "no molecules represented by NaCl. The equality in number of sodium and chlorine atoms is arrived at by a chess-board pattern of these atoms; it is a result of geometry and not of a pairing-off of the atoms" as "repugnant to common sense" and "not chemical cricket".

enabling the Gulf Stream to convey an enormous amount of energy from the tropics to Northern Europe and keeping the temperature of the seas around Cape Horn low, even in summer.[4] Increasing the motion of the molecules accounts for only half of the liquid's heat capacity; there must be a structure in the liquid which the rest is utilised in deforming.

Intermolecular bonding of this sort occurs in compounds where hydrogen can lie between two highly electronegative fluorine, oxygen or nitrogen atoms. The character of this hydrogen bonding remains somewhat obscure. Traditionally, writers have been reluctant to question a fixed valency of one and ascribe hydrogen two covalent bonds. It remains controversial whether it is a purely electrostatic interaction due to a build up of positive charge on the hydrogens and negative charge on the electronegative element (oxygen in water), or has a significant covalent contribution. But it is important to understand that hydrogen-bonded water constitutes a cooperative interaction, in which the interaction of a given water molecule with an already existing cluster of hydrogen-bonded molecules is more favourable, and therefore more probable, than the interaction with another single molecule to form a dimer (Franks 2000, p. 14; Needham 2013, p. 59). Hydrogen bonding therefore disposes water molecules to amalgamate in very large structures, discouraging the existence of even dimers (i.e. pairs of molecules), and certainly monomers (single, uncombined molecules).

Water molecules aggregate in inter-molecular species which might be represented by $(H_2O)_2$, $(H_2O)_3$, …, $(H_2O)_n$, …. But these are not permanent species, nor even stable over nanoseconds (10^{-9} s). The situation is a dynamic one, in which clusters are continually being formed and broken, reaching equilibrium under stable conditions in the sense that their concentrations remain constant because the rates of formation and destruction are equal. The dynamic equilibria can be represented as follows:

$$\begin{aligned} 2H_2O &\rightleftharpoons (H_2O)_2, \\ H_2O + (H_2O)_2 &\rightleftharpoons (H_2O)_3, \\ &\cdots \\ H_2O + (H_2O)_n &\rightleftharpoons (H_2O)_{n+1}, \end{aligned}$$

The double arrows indicate a state of equilibrium in which the concentrations of the species on each side of the "equation" remain constant. It should be stressed, however, that there is great uncertainty about how many such species there are and what their actual structure is. Much of the present view is derived by inference from macrofeatures along the lines indicated above. Even so, enough is known to motivate a modification of this scheme.

[4]Latent heats of vaporisation are often higher than latent heats of fusion, and heat capacities of the liquid are often greater than those of the solid and gaseous phases of the same substance. But in the case of water the differences are more marked.

A second series of equilibria is established arising from the disruption of the intermolecular covalent bonding with dissociation of water into ions—positively charged hydrogen cations and negatively charged hydroxyl anions:

$$H_2O \rightleftharpoons H^+ + OH^-.$$

These ions are not free particles but attach to the hydrogen-bonded clusters, producing equilibria of the following kinds:

$$3H_2O \rightleftharpoons H_3O^+ + (H_2O)OH^-,$$
$$\ldots$$
$$(2n+1)H_2O \rightleftharpoons (H_2O)_nH^+ + (H_2O)_nOH^-.$$

In pure water the hydrogen and hydroxyl ion concentrations[5] are each very small, 10^{-7} g ions per litre. It might be thought that such tiny amounts could be ignored. But the dielectric constant of water depends on the existence of these ions and how they are hydrated by attaching to clusters. This concentration determines the neutral pH value (defined as the negative logarithm of the hydrogen ion concentration) of 7. Acidity is the phenomenon of the pH value decreasing (i.e. hydrogen ion concentration increasing) in an aqueous solution (homogeneous mixture of one or more substances with water). These ionic species, although in very low concentrations, account for some of water's distinctive properties. The pH of the ambient medium controls whether haemoglobin holds onto or releases oxygen, for example, and so is a matter of life and death for us. If microscopic species determine the characteristic features of a substance, then they can't be ignored just because there are, on average, relatively few of them.

Although this discussion has concentrated on water in view of its prominence in philosophers' discussions of chemical substances, it should be borne in mind that there are many substances which exhibit non-molecularity in one way or another. Alcohols, for example, are like water in their association and disassociation behaviour, and carbon tetrachloride comprises temporary clusters in the liquid state. See Needham (2013) for an overview of the various guises in which hydrogen bonding appears and how it is distinguished from related forms of intermolecular attraction.

Derivations of the phase rule such as those given in Sect. 6.4 take no account of the various microparticles present in the components. It is a theorem of classical thermodynamics, which is a purely macroscopic theory, neutral with respect to microscopic interpretations. In order to relate macro- and microdescriptions, a standard move in chemical theorising is to consider thermodynamic equilibria, representing stable and macroscopically unchanging states of affairs, to correspond to dynamic equilibria on the micro-scale. On the latter conception, there is a

[5]Including all the hydrated hydrogen ion species in the hydrogen ion concentration, and similarly for the hydroxyl ion concentration.

continual removal of reactants as they combine to yield products, whilst at the same time reactants are continually replenished as the products re-combine, and a dynamic equilibrium is established when the rate of the forward reaction is the same as the rate of the reverse reaction. Thus, the dissociation of water into hydrogen ions and hydroxyl ions proceeds, whilst the recombination of these ions occurs, removing ions after they have been created. If the forward reaction were allowed to proceed alone, by somehow removing the ions when they are formed, then the reaction would proceed until all the water is dissociated. But this doesn't happen and a balance is reached in which the rate of dissociation equals the rate of association, when the concentration of undissociated species remains constant, despite the continual dissociation, because dissociated species are continually replaced. The concentration of dissociated species remains constant for the same reason.

The word "species" is often used in discussions of the phase rule to indicate kinds whose number exceeds c, the number of independent components, and this includes microscopic species. From the point of view of the phase rule, then, the dynamic equilibria between the various cluster species H_2O, $(H_2O)_2$, $(H_2O)_3, \ldots,$ $(H_2O)_n, \ldots$ in liquid water involving association by hydrogen bonding are chemical reactions at equilibrium. Accordingly, if there are n species, whatever n might be, there are $n - 1$ equilibria, and therefore just one independent substance. So although chemists are still pursuing unanswered questions about these clusters, they are sure that they are in dynamic equilibrium with one another, and the exact number is immaterial to the question of how many substances are present. Water is a single substance. The dissociation into ions may seem to present a different case in so far as the equilibrium $H_2O \rightleftharpoons H^+ + OH^-$ involves 3 species. But this one reaction yields the further constraint that the concentrations of hydrogen and hydroxyl ions are equal (in molar units), maintaining electrical neutrality. So the three species are matched by two constraints, which brings the number of substances back to one. And as before, the ionic cluster complexes introduce $2n + 1$ species along with $2n$ constraints, which again fails to raise the number of independent substances above one. This explains the insensitivity of the thermodynamic criterion of purity based on the phase rule to the variety of particles in water.

What appears macroscopically as an unchanging state of affairs is thus the scene of rapid change at the microlevel. Although the changes balance to maintain steady concentrations when dynamic equilibrium is established, the fixed concentrations are not constituted by fixed quantities of matter. The half-life of the dissociation of water (time for half of a given amount of water to decompose) is about 20 ms (20×10^{-6} s). So after a very short time on the human scale, say half a second, an appreciable proportion of the water molecules in any given quantity of water will have been dissociated and the ions from any particular molecule will have recombined, though not necessarily with one another. Similarly, each of the steps in the formation and dissociation of polymeric clusters of water species is established in water in the liquid state within intervals of time not exceeding 10^{-11} s. Accordingly, the quantity of matter which is such a polymeric species is in general so for a time of this order of magnitude and no longer. For a significantly longer time, say of the order of a microsecond (10^{-6} s), this quantity may well not be such a cluster, but a sum whose parts form parts of various such clusters for different subintervals of the time in question. Similarly, a part of such a cluster which is a water molecule at some time will not be an H_2O molecule for very long. Given what was said about the dissociation half-life of H_2O, the probability of a quantity of matter being an H_2O molecule for times of macroscopic orders of magnitude becomes quite small. Accordingly, a quantity which is an H_2O molecule for an appropriately short time will often be a sum of disparate and dispersed parts which don't constitute an H_2O molecule for times some factors of 10 longer.

How does all this bear on the question of whether the distributive condition applies? Quine's original objection was that "water" is not distributive because it doesn't apply to parts too small to count. It applies "down to single molecules but not to atoms" (Quine 1960, p. 98). But the issue is not quite so cut and dried. Taking the dyadic character of the "water" predicate into account, suppose "$Water(\pi, t)$" is true, where π is a macroscopic quantity and t a macroscopic time. Considering ever smaller subintervals of t, some subinterval t' is the lifetime of a particular H_2O molecule. It doesn't follow that the remainder of π is susceptible to a partition into H_2O molecules at t', and it is unlikely in the extreme that the π for which $Water(\pi, t')$ is true can be resolved into a mereological partition of molecules at t'. All the molecules do not dissociate into ions in step, as this would require. While some molecules survive intact throughout the period, others will dissociate, and ions will associate to form new molecules during the same period. This remains true for a t' which is the shortest lifetime of all the molecules appearing in π during t.

A molecule in the liquid isn't like a water molecule in isolation (in the gas phase at very low pressure), but will be subject to all the stresses imposed by the ambient medium. In fact, even very much larger parts of water, although small enough to be at the nano scale, behave differently when isolated from when part of bulk water. According to recent investigations into the process of ice crystallisation, the frequency of ice nucleation varies with the volume of water, raising the question of whether water at the nanoscale can still be regarded as equivalent to bulk water. Li et al. (2013) show that the ice nucleation rate at the nanoscale can be several orders of magnitude smaller than that of bulk water, suggesting that water at this scale can no longer be considered bulk water and that the boundary for such deviant behaviour (size of the droplet where the difference vanishes) varies with temperature. These experiments were conducted on very small drops of water, isolated from bulkier quantities of water, as distinct from examinations of the behaviour of very small parts immersed in bulkier quantities of water. It raises the issue of the difference between parts in contact with the whole and parts removed from the remainder of the whole familiar from the discussion of what Holden says about the potential parts doctrine (Sect. 5.7). Isolated nano-sized quantities of water might well behave differently from larger isolated quantities, just as water in brine does. The observations of Li et al. therefore don't necessarily provide any insight into how the line should be drawn between the micro-composition and the macroscopic parts of a connected quantity of matter.

It might well be said that some such considerations bear on Quine's claim that a quantity is water "down to single molecules". Even for a molecular substance like methane, the molecules are said not to display characteristic properties of methane such as its melting and boiling point, which call for macroscopic-sized quantities. But although it is true that isolated methane molecules don't have melting and boiling points, it is not so clear that methane molecules immersed in a substantial quantity of methane, whose properties are generally dependent on the surrounding medium, don't. And in the absence of reasons for disregarding properties such as these that are normally regarded as characteristic of the substance, the molecules must surely be regarded as being of the substance kind when making their contribution to the exhibition of macroscopic properties in the context of the bulk material during the time in question. (N.B. the distributive condition speaks of parts of a whole quantity *during the time when* the whole possesses the substance property in question.) Molecules of a different substance dissolved in methane wouldn't contribute in the same way, but lower the melting point—i.e., the temperature at which macroscopic quantities would transform from solid to liquid at normal pressure—in relation to the melting point of pure methane. So in specifying the dispositional property of melting at $-183\,^{\circ}C$, to the effect that a quantity of methane would be liquid if brought to this temperature at standard pressure and supplied with the latent heat of fusion, we must add "in the absence of any other dissolved substance". Why shouldn't it also be understood to include the condition "when immersed in a sufficiently large connected quantity of methane"?

A similar point applies to the atoms Quine suggests are parts of water molecules that are not water. The modern idiom speaks of hydrogen and oxygen in water and of hydrogen and oxygen atoms in water molecules. But just as there is no inflammable gas in water, there are no hydrogen atoms with electronic configuration $1s^1$ characteristic of isolated hydrogen atoms or oxygen atoms with electronic configuration $1s^2 2s^2 2p^4$ characteristic of isolated oxygen atoms. (Likewise, in methane molecules there are no carbon atoms with the electronic configuration $1s^2 2s^2 2p^2$ characteristic of isolated carbon atoms in the ground state.) There are hydrogen and oxygen nuclei in water molecules. Hydrogen nuclei (protons) might be likened to hydrogen ions, which are just as much particles constituting water as are water molecules, constituting, like water molecules, transient components of larger polymeric species. But there is more to an atom than its nucleus. Although the electrons do not contribute much to the mass of the atom, they are essential to its overall electrical neutrality and the chemical bonds which it forms. Electrons are present in the water molecules in numbers corresponding to those in two isolated hydrogen atoms and one oxygen atom. But even on the Lewis theory of the pairing of electrons in covalent bonds, there is no assigning some of the electrons in the water bonds to a unit comprising just an oxygen atom and others to units each comprising just hydrogen atoms. The quantum mechanical feature of the indiscernibility of electrons drives the point home, in view of which we have to rest content with the statement that there are no entities like the isolated atoms just described in water. Whatever else is entailed by the claim that there are hydrogen and oxygen atoms in water molecules, the mereological claim that there are atoms forming parts of water molecules is not true. There are at best sums of eight neutrons, eight protons and eight electrons which are parts of (what are sometime) water molecules, but none of them is, or constitutes, an oxygen atom while the quantity they are part of is a water molecule. There are neither hydrogen nor oxygen atoms in water molecules.

So what of submolecular parts of sometime molecules of water? Do they provide counterexamples to the distributivity of the "water" predicate? For any macroscopic quantity π and macroscopic time t for which "$Water(\pi, t)$" is true, π is a sum of fleeting components of larger molecular and polymeric species which lacks the structure of a molecular partition and doesn't freeze into such a partition for shorter intervals. It is the same π that $Water(\pi, t')$ is true of, however small the subinterval t' of t. If nano-scale parts of water are water even though they might be considered not to be when isolated from bulk water, as I think it reasonable to claim, is it not also reasonable to claim that all parts of π, no matter how small, are water for all parts of the time when π is water? At all events, the picture Quine paints to motivate denying the claim is not applicable.

The situation is similar to the case described in the last chapter where water is heated to 2000 K. In the gas phase, water dimers, not to mention larger clusters, rapidly become more of a rarity as the temperature rises, so that at normal pressure, water is to all intents and purposes molecular for temperatures ranging from a little above the boiling point until it starts to dissociate around 500 °C. But at higher temperatures, because of the dynamic equilibrium at the microlevel, a quantity which is all water under such conditions for macroscopic lengths of time will not be partitioned into water molecules. It will be a sum of sometime parts of water molecules, sometime parts of dihydrogen molecules and sometime parts of dioxygen molecules. Parity of reasoning would suggest that the two situations should be treated in the same way: if liquid water is a single substance despite the heterogeneous microstructure, then the gas at high temperatures is also a single substance despite the dissociation. Ammonia at high temperatures will similarly be a sum of sometime parts of ammonia molecules, sometime parts of dinitrogen molecules and sometime parts of dihydrogen molecules. Liquid ammonia also exhibits hydrogen bonding promoting the association of molecules, although it cannot form the extensive cross-linked network that is found in water because each molecule cannot be simultaneously linked to others by so many hydrogen bonds. It also dissociates into positive ammonium and negative amide ions:

$$2NH_3 \rightleftharpoons NH_4^+ + NH_2^-.$$

Summarising, if macroscopic criteria based on the phase rule determine the applicability of the substance predicate "water", then it holds of sufficiently large quantities for sufficiently long times over which the microscopic fluctuations are smoothed out. It doesn't apply to what microdescriptions are true of when considered in isolation from the macroscopic bulk, for much shorter times. There are no such parts of macroscopic quantities of water. For these macroscopic times, a macroscopic quantity of water will be a sum of scattered fragments of microparticles which cannot be partitioned into particles such as molecules. Even where the time for which "water" applies to a macroscopic quantity is short enough for descriptions of microscopic particles to apply, the quantity won't in general be partitioned into particles because the dissociations and associations won't be in step. Arguably, microscopic parts of water are water. Their features will vary considerably over macroscopic times as they interact in the course of dissociations and associations of molecular species. But even for much shorter times of the order of lifetimes of molecular species in liquid water, these microparts don't display properties they would if isolated. Their properties are characteristic of being part of what definitely is water and might therefore reasonably be regarded as water for the times in question.

In general, then, larger macroscopic quantities of matter sustain macroscopic substance properties for sufficiently long times, and very small parts of macroscopic quantities of matter sustain structural properties of microscopic entities present in bulk matter for times of a microscopic order of magnitude, but not longer. This is not independent of the prevailing conditions and the kind of substance at issue, however. The molecules in a quantity of water confined entirely to the gas phase under conditions approaching those of an ideal gas (obeying the Charles-Boyle gas law to a very good approximation, requiring an appropriate temperature and low pressure) are so thinly distributed that they hardly interact. Under such conditions the molecules have a considerably greater longevity than molecules in the liquid phase. But as we will see in the next section, this is not so if the gas is in contact and at equilibrium with the liquid (in a multiphase system).

7.3 What Do Phase Predicates Apply To?

How do the general considerations raised in the previous section apply to the other broad category of macroscopic kinds, namely phases? A dynamic balance between rates of opposing processes is also at work in the equilibria between phases in a complex mixture, which undergo exchange of material. The equilibrium between liquid and vapour, for example, involves particles leaving the liquid surface and entering the vapour whilst being replaced by other particles from the vapour entering the liquid. Again, in the space of what, on the human time scale is an exceedingly short interval of a fraction of a second, substantial parts of the matter in the liquid will have gone into the vapour phase to be replaced by different matter coming in from the vapour phase. There is a continual interchange of material between the two phases. This is often put by saying that water molecules in the vapour phase will enter the liquid phase, be subject to the interactions and become integrated into the microstructure of the liquid. The thought here would be less ambiguously expressed, however, by saying that what are water molecules in the vapour phase will become integrated into the microstructure of the liquid, dropping any suggestion that the molecular units remain intact throughout the course of this process.

On a macroscopic time scale, what is at one time, say, liquid in a liquid-vapour system, or solid water at the triple point where solid, liquid and gas are all at equilibrium, is in general not all liquid (solid) at another. This raises the question of what it means to say that some quantity is in some particular phase, say liquid, for some time (at some interval). It can't be given the same mereological interpretation suggested above for the macroscopic substance predicate "water". If "is liquid" were distributive, i.e. any part of a quantity which is liquid at a time is liquid at any subinterval of that time, then for liquid at equilibrium with another phase, the longer the time, the smaller would be the quantity counted as liquid. In the long run, nothing would be liquid for a time during which all liquid is exchanged with the other phase. But even before reaching this extreme, imposition of the distributive condition would mean that whatever chunk of matter satisfies the condition for a given time is not discernible by perception, being some quantity occupying a proper part of the region apparently occupied by liquid. This raises the question of what, if anything, is discerned by perception for a given time and apparently referred to as "the liquid".

Alternatively, "is liquid" might be interpreted accumulatively, by analogy with the way the region occupied by a body (quantity or individual) at a time is understood to be the region it sweeps out during that time (Sects. 2.3 and 2.4) and with the accumulative interpretation of the material constitution of individuals (Sect. 3.2). Here the understanding is that, just as some part of the region swept out during t might not be occupied for some subinterval of t, so some part of the quantity which is liquid for t would not be liquid for some subinterval of t. The longer the time, the greater would be the quantity counted as liquid. As for the formulation of the summation principle, it would be incorrect to say simply that a quantity of which "is liquid" is true for time t is the sum of quantities which are liquid for some subinterval of that time because this sum would exceed π and include all the world's liquid at some subinterval of t. Accordingly, the sum is restricted to parts of π:

$$Liquid(\pi, t) \supset \pi = \Sigma \pi' \subseteq \pi \, \exists t' \subseteq t \, Liquid(\pi', t').$$

A quantity is always the sum of its parts, but this is not that trivial theorem because it says that the sum of only some of π's parts, namely those liquid at some subinterval of t, is in fact identical with π. It is not true that every part of π is liquid for *some* subinterval of t since some parts of what is liquid at time t are sums of two or more parts that are liquid at disconnected times, in which case there is no (connected) time when the sum is liquid.

The first proposal, based on the distributive condition, would count as liquid a quantity of less mass than what apparently is the mass of the liquid from the purely macroscopic perspective, and the longer the time, the smaller the quantity counted liquid. For a sufficiently long time, there would be no liquid in a closed system. The second proposal, based on the accumulation condition, would count as liquid a quantity of greater mass than the apparent mass of the liquid from the purely macroscopic perspective, and the longer the time, the greater the quantity counted liquid. For a sufficiently long time, all the water in a closed system would be liquid. But then none of the water would be any other phase, say gas, for this time. For a closed system with a single substance exhibiting several phases, one phase might be interpreted accumulatively and the others distributively. They couldn't all be interpreted accumulatively.

There may well be a certain fixed volume of liquid in a closed container enclosing a single substance exhibiting liquid and gas phases at equilibrium over a certain period. From the macroscopic perspective, it seems natural to take it that one and the same thing makes up this volume throughout the period in question. But the above considerations show that there is no such thing. Definite quantities are defined by the distributive condition and by the accumulation condition, either of which might serve as a referent of the liquid, but the fixed volume cannot be attributed to either of these. A third course of delimiting a definite quantity by reducing the time interval to an instant might suggest itself. But this would take us further than the limit imposed by statistical fluctuations, beyond the limit set by the time-energy uncertainty principle, and into a realm where scientists hardly dare speculate. This might not deter the metaphysician content to imagine water at the microlevel arranged "waterwise", but others might be sceptical of attempts to construe this as anything more than an empty gesture. In any case, the issue would still arise of what is the liquid for finite times, calling for some integrating construction over the different instantaneous objects.

There always being something that bears the phase predicate, no matter how long the time at issue, counts perhaps in favour of the accumulative interpretation for one phase. Whereas the distributive interpretation implies that every part of what is liquid at time t is liquid for *every* subinterval of t, the accumulative interpretation doesn't even require that every such part is liquid for *some* subinterval of t. As noted above, some parts of what is liquid at time t might be sums of two or more parts that are liquid at disconnected times, in which case there is no (connected) time when the sum is liquid. This suggests that, on the accumulative interpretation, the antecedent of the cumulative condition,

$$\forall \pi' \subseteq \pi \, \forall t' \subseteq t \, \exists \pi'' \subseteq \pi' \, \exists t'' \subseteq t' \, Liquid(\pi'', t'') \supset Liquid(\pi, t),$$

is not fulfilled. It is not the case that every part of what is liquid at time t always has some part that is liquid for some subinterval. Part of the liquid that goes into the gas phase sometime during the time t at issue may well never go back into the liquid phase during t.

Consider a somewhat more complicated case with several substances present, the calcium carbonate equilibrium described in Sect. 6.4. In a standard monograph on the phase rule, Ricci (1951, p. 10) says of this case, "With CaO and CO_2 [as independent components], the phase $CaCO_3$ consists of CaO + CO_2, with $CaCO_3$ and CaO [as independent components], the phase $CO_2 = CaCO_3 - CaO$." Although he leaves the reader to infer the significance of the equations, it is surely natural to interpret "+" and "−" as mereological operations on quantities, and thus to see the chemical formulas as predicates expressing properties borne by these quantities. But there is clearly some ambiguity about whether substance or phase properties are at issue, which needs to be resolved if the algebraic interpretation is to be at all clear.

It would seem that the solid is separate from the gas (CO_2) phase, and the description of the calcium oxide phase as confined to isolated "islands" would seem to suggest that it is mereologically separate from the calcium carbonate phase.

But how, then, is Ricci's statement that "the phase $CO_2 = CaCO_3 - CaO$" to be understood? If the calcium carbonate and oxide phases are mereologically separate, then the mereological difference of the calcium carbonate phase less the calcium oxide phase is just the calcium carbonate phase. Similarly, if the CO_2 phase is separate from the $CaCO_3$ phase, the mereological difference of $CaCO_3$ less CO_2 is just $CaCO_3$. So if the operations are applied to the phases, as Ricci clearly suggests, it seems there is no matter constituting the gas phase! Similarly, the $CaCO_3$ phase will not consist of the CaO phase + the CO_2 phase if "+" is mereological summation. How should we get a viable interpretation out of this?

If the calcium carbonate equilibrium involves two independent *substances*, and these may be chosen in different ways, then the alternative choices must divide the total quantity of matter into substances in essentially the same way. All three substances stand in certain mereological relations to one another, and any two can be related to a third in terms of their relationship to one another. On the choice of calcium carbonate and calcium oxide as independent substances, the suggestion was that carbon dioxide can be taken as the calcium carbonate less the calcium oxide. If "less" is to be understood as mereological difference, and the carbon dioxide is not a null entity but some definite quantity of matter distinct from both the calcium carbonate and the calcium oxide, we are naturally led to the following interpretation. The entire quantity of matter in the equilibrium is all calcium carbonate, and the calcium oxide is a proper part of this. The mereological difference between the calcium carbonate and the calcium oxide is then defined, and is another proper part of the calcium carbonate separate from the calcium oxide which we may call carbon dioxide.

An alternative choice of independent substances would be calcium carbonate and carbon dioxide, taking the calcium oxide to be the calcium carbonate less the carbon dioxide and recovering the same overall mereological structure as before. The calcium oxide, like the carbon dioxide, is a proper part of the calcium carbonate, which is the entire quantity of matter in the system. A third choice of independent substances makes a mereological partition of the entire quantity of matter, the calcium oxide and the carbon dioxide being separate and their sum is the entire quantity. In this case, the calcium carbonate is the mereological sum of the calcium oxide and the carbon dioxide, which gives us a mereological interpretation of the "+" in the above quotation from Ricci. But for the reasons just explained, it is not, as Ricci says, the calcium carbonate phase which is this sum, but the calcium carbonate substance.

Phases can't be interpreted in the same way as substances. The two predicates "calcium carbonate substance" and "calcium carbonate phase" are not coextensive. Whereas the predicate "calcium carbonate substance" applies to the entire quantity of matter, the phase predicate "calcium carbonate phase" apparently doesn't. As we saw above, what it applies to is apparently separate from both the calcium oxide phase and the carbon dioxide phase. But on the van't Hoff principle connecting macro- and microphenomena, according to which a static equilibrium at the macrolevel corresponds to a dynamic equilibrium at the microlevel in which a balance obtains between the rates of counteracting processes, matter moves between the phases. In the present case, that occurs at the same time as equilibrium reigns between dissociation and association reactions. But like the two-phase water case, the phase predicates describing the situation don't apply to a fixed quantity throughout an interval. Matter is constantly exchanged between the phases, so that whilst some of the calcium carbonate phase matter becomes calcium oxide phase and gas (carbon dioxide) phase, other matter that is calcium oxide and gas phase becomes calcium carbonate phase. Parts of each of the substances combine to temporarily form one of the solid phases, whilst parts of this solid phase dissociate and become part of the gas phase and part of the other solid phase. The partitioning of the original quantity into calcium oxide and carbon dioxide substance is permanent (for the duration of the equilibrium), but it doesn't coincide with the division into what the calcium oxide and carbon dioxide phase predicates apply to for a given time. This latter division is neither exhaustive, mutually exclusive nor fixed for the duration of the equilibrium. This accommodates the fact that the substance predicates "calcium carbonate substance", "calcium oxide substance" and "carbon dioxide substance" are related to the number c in the phase rule, whereas the phase predicates "calcium carbonate phase", "calcium oxide phase" and "carbon dioxide phase" are counted by the number f in the phase rule.

7.4 Indiscernible Particles

Time, at least when considered in abstraction, would seem to be infinitely divisible: as expressed by axiom MT10 of Sect. 1.3, every time has a proper part. It was suggested in the last section but one that restrictions on the distributivity condition applied to substance predicates are not so obviously true as has seemed to some authors: parts of a macroscopic quantity of water such as water molecules and their submolecular parts at short subintervals of the time when the quantity is water might reasonably be considered water too. This is to consider all such putative parts as entities of essentially the same kind as the macroscopic quantities which uncontroversially do bear the property of being water at some time, i.e. as being macroscopic quantities themselves. But the suggestion was raised at the beginning of that section that microentities might be ruled out as parts of macroscopic quantities in virtue of belonging to the quantum domain. I return to that suggestion now.

Can a dichotomy be drawn between macroscopic entities (quantities) and microscopic entities by distinguishing microscopic entities on the basis of some characteristic feature and excluding them from the domain of parts of quantities? Indistinguishability is a characteristic feature universally ascribed to quantum mechanical particles that distinguishes them from macroscopic entities. Messiah and Greenberg (1964) give the classical formulation of the indistinguishability postulate as the unobservability of any difference between a wave function describing a state and the wave function resulting from the permutation of particle coordinates. The following passage of Penrose's gives a sense of what is involved:

> According to quantum mechanics, any two electrons must necessarily be completely identical, and the same holds for any two protons and for any two particles whatever, of any particular kind. This is not merely to say that there is no way of telling the particles apart; the statement is considerably stronger than that. If an electron in a person's brain were to be exchanged with an electron in a brick, then the state of the system would be exactly the same state as it was before, not merely indistinguishable from it! The same holds for protons and for any other kind of particle, and for the whole atoms, molecules, etc. If the entire material content of a person were to be exchanged with the corresponding particles in the bricks of his house then, in a strong sense, nothing would have happened whatsoever. What distinguishes the person from his house is the pattern of how his constituents are arranged, not the individuality of the constituents themselves. (Penrose 1989, p. 32, quoted by French and Krause 2006, p. 263)

The spirit, but not the terminology, of this passage is followed here. Microentities are said here to be indiscernible, and "identity" is reserved for a relation between macroscopic entities. An idea inspired by Newton da Costa and developed by French and Krause (2006) is to treat the two kinds of entity on the basis of a so-called Schrödinger logic, which treats identity statements between variables ranging over microscopic entities as not well formed and therefore meaningless.[6]

The basic first-order theory of mereology, formulated in terms of variables $\pi, \rho, \sigma, \ldots$ understood to range specifically over macroscopic quantities (Sect. 1.2), is expanded into a two-sorted system as a "Schrödinger" logic with identity statements defined (well-formed) only for macroscopic variables. Microscopic variables $m, n, o, \ldots$ ranging specifically over microentities are added, for which identity statements are ill formed. General metalinguistic variables $u, v, w, \ldots$, taking the place of either micro- or macroscopic variables, are used in the formulation of schemas. A primitive relation of indiscernibility, $\approx$, forming well formed sentences with both kinds of variable, satisfies axioms making it an equivalence relation.

Quantities are said to be *composed* of microentities, expressed with a new dyadic relation $\mathbb{C}uv$, read "u composes v", satisfying a fundamental principle:

NoMacmic $\quad \forall\pi \sim\mathbb{C}\pi v$

restricting the relation $\mathbb{C}$ to a relation standing between microentities on the one hand, and either micro- or macroscopic entities on the other. Quantities don't compose anything. A molecule, let us suppose, is a microentity that is composed of other microentities, and composes quantities, but no quantity composes any microentity.

[6] An alternative interpretation of the insensitivity of quantum mechanics to the permutability of particle has it that, although these particles are not discernible by their properties, they are discernible by certain relations holding between them, in view ow which there is an irreflexive and symmetric relation of weak discernibility standing between them (Saunders 2006), The intention here is not to adjudicate between this and French and Krause's interpretation. That would involved too large a digression in a book primarily concerned with macroscopic objects. The purpose is simply to explore one interesting possibility of restricting the scope of the part relation.

It would seem that the principle of the insensitivity to permutations expressed by Penrose in the above passage could be captured in mereological terms by a thesis along the following lines[7]:

$$n \approx o \ \wedge \ \mathbb{C}nm \wedge \ {\sim}\mathbb{C}om \ .\supset \ (m - n) \cup o \approx m.$$

A quasi-mereology for microentities with the composed relation and operations of difference and sum is outlined in this section, allowing for the formulation and adoption of such a principle of the insensitivity to permutations.

The point of the present exercise is to ensure that, although microentities compose quantities, they are not parts. A similar policy has already been instigated in the present many-sorted framework where mereological features are laid down separately for each of the categories of times, regions and quantities with no cross-category relations or operations. Expression such as $t \mid \pi$, $p \mid \pi$ and $t \mid p$ have been implicitly regarded as ill formed, without any import on the mereological principles governing the entities in these categories. This strategy is naturally extended to the new category of microscopic entities by not allowing expressions of the kind $m \mid \pi$, $\pi \mid m$, $t \mid m$, $m \mid t$, $p \mid m$ or $m \mid p$, involving microscopic variables in the standard relations of mereology, as well formed. An alternative strategy would be to allow expressions of the kind $m \mid \pi$ and $\pi \mid m$ to be well formed and to lay down a general axiom that microentities are always separate from entities of any kind. But this course is not followed here.

Composition and parthood are related by the principle

micMac $\quad \mathbb{C}m\pi \ \wedge \ \pi \subseteq \rho \ .\supset \ \mathbb{C}m\rho,$

which is a natural extension of transitivity of parthood linking the micro and macro realms. This includes the special case

$$\mathbb{C}m\pi \ \wedge \ \pi = \rho \ .\supset \ \mathbb{C}m\rho.$$

Although lack of the usual identity criterion may discount composition from being a proper mereological relation, a quasi-mereology can be developed for microentities with a view to expanding on the principles governing the composition relation by analogy with parthood. To facilitate comparison with standard mereology as developed here, the quasi-mereological primitive is chosen to be micro- or quasi-separation, $\wr$, alongside the new primitive $\approx$, and composition, $\mathbb{C}$, is defined by

Df $\mathbb{C}$ $\quad \mathbb{C}vw \ \equiv \ \forall o (o \wr w \supset o \wr v).$

Instead of the antisymmetry property as it applies to parthood, the following analogue is laid down as an axiom

IndistAx $\quad \mathbb{C}mn \wedge \mathbb{C}nm \ .\supset \ m \approx n.$

[7] Cf. Theorem 26 of French and Krause (2006, p. 296).

IndistAx serves as a criterion of indistinguishability, much like the mereological criterion of identity MQ1. Reflexivity and transitivity of composition,

$$\text{Crefl} \qquad \mathbb{C}mm$$
$$\text{Ctrans} \qquad \mathbb{C}mn \wedge \mathbb{C}no \;.\supset\; \mathbb{C}mo,$$

follow directly from Df $\mathbb{C}$. $\mathbb{C}mm$ is a theorem of predicate logic when written out in accordance with Df $\mathbb{C}$. For transitivity,

$$\begin{aligned}\mathbb{C}mn \wedge \mathbb{C}no \;.\equiv.\; & \forall o' (o' \wr n \supset o' \wr m) \wedge \forall o' (o' \wr o \supset o' \wr n) \\ .\equiv\; & \forall o' ((o' \wr o \supset o' \wr n) \wedge (o' \wr n \supset o' \wr m)) \\ .\supset\; & \forall o' (o' \wr o \supset o' \wr m),\end{aligned}$$

which is $\mathbb{C}mo$ as required.

The analogue of overlap, $\odot$, is defined by the existence of a common microcomponent:

$$\text{Df}\,\odot \qquad m \odot n \;\equiv\; \exists o\, (\mathbb{C}om \wedge \mathbb{C}on),$$

which entails the symmetry and reflexivity of $\odot$. The relation is governed by the following analogue of the axiom MQ2 of standard mereology,

$$\text{SepAx} \qquad m \odot n \;\equiv\; \sim\!(m \wr n),$$

which in turn entails the symmetry and irreflexivity of $\wr$. Two quantities, or a quantity and a microentity, might microoverlap (be microseparate) in virtue of a common (no common) microcomponent. Composition and quasi-separation are simply related by the theorem

$$\text{T1} \qquad u \wr v \wedge \mathbb{C}mv \;.\supset\; u \wr m.$$

For suppose not, i.e. $u \wr v$, $\mathbb{C}mv$ and $u \odot m$. Then there is a microentity which composes both u and m, and so both u and v by $\mathbb{C}$trans, contradicting $u \wr v$.

Proper composition, $\overline{\mathbb{C}}$, is defined by the schema

$$\text{Df}\,\overline{\mathbb{C}} \qquad \overline{\mathbb{C}}vw \;\equiv.\; \mathbb{C}vw \wedge \sim\!\mathbb{C}wv.$$

This definition avoids the use of identity as it featured in the definition of proper parthood. But the definition clearly entails that proper composition is irreflexive (on pain of contradiction) and asymmetric, and by IndistAx and Df. $\overline{\mathbb{C}}$,

$$\text{T2} \qquad \mathbb{C}mn \;\supset.\; \overline{\mathbb{C}}mn \vee m \approx n.$$

Transitivity is established by a straightforward argument. Assume $\overline{\mathbb{C}}mn \wedge \overline{\mathbb{C}}no$. Then clearly $\mathbb{C}mo$, and it remains to show $\sim \mathbb{C}om$. Assume not, i.e. $\mathbb{C}om$. Then $\mathbb{C}nm$ (given $\mathbb{C}no$ by $\mathbb{C}$trans), contradicting $\overline{\mathbb{C}}mn$ which implies $\sim\!\mathbb{C}nm$.

Proper composition should naturally entail distinguishibility (the negation of indistinguishability). Composite microentities such as sodium atoms, for example, though indistinguishable from one another, are distinguishable from their microcomponents. This entailment would follow from Df. $\overline{\mathbb{C}}$ and the converse of IndistAx.

IndistAx is the analogue of the mereological criterion of identity, MQ1, the converse of which follows from the laws of identity. No such principles are available to demonstrate the converse of IndistAx, and this seems an unwarrantedly strong principle to accept in justification of this criterion of distinguishability. The criterion of distinguishibility is therefore taken as axiomatic:

DistAx $\quad \overline{\mathbb{C}}mn \supset \sim(m \approx n)$.

In view of the NoMacMic principle, $\mathbb{C}m\pi \supset \overline{\mathbb{C}}m\pi$ and $\sim\overline{\mathbb{C}}\pi m$ are theses.

The litmus test of the strength of the analogy with mereology is whether composition, as interpreted here, sustains the analogue of the strong supplementation principle (cf. Simons 1987, p. 29):

$\mathbb{C}$SS $\quad \sim\mathbb{C}mn \supset \exists o\,(\mathbb{C}om \wedge o \wr n)$.

A photon, *n*, is not composed of an electron, *m*, for example. This principle would suggest that there must be some microentity, *o*, of which the electron is composed that is microseparate from the photon (with apologies for the misplaced definite articles). This wouldn't be the case if the relation in the first existentially quantified conjunct were proper composition. But as it stands, the relation is composition, which is reflexive, so the consequent is satisfied by the electron itself, which is microseparate from the photon. $\mathbb{C}$SS can be demonstrated as follows:

$$\begin{aligned} \sim\mathbb{C}mn &\equiv \sim\forall o\,(o \wr n \supset o \wr m) && [\text{Df}\,\mathbb{C}] \\ &\equiv \exists o\,(o \wr n \wedge \sim o \wr m) \\ &\equiv \exists o\,(o \wr n \wedge \exists o'\,(\mathbb{C}o'o \wedge \mathbb{C}o'n)) && [\text{SepAx, Df}\,\odot] \\ &\equiv \exists o\,\exists o'\,(o \wr n \wedge \mathbb{C}o'o \wedge \mathbb{C}o'n) \\ &\supset \exists o'\,(n \wr o' \wedge \mathbb{C}o'n) && [\text{symm of}\ \wr, \text{T1}], \end{aligned}$$

as required. $\mathbb{C}$SS holds when the variable "*n*" is replaced by a macroscopic, quantity variable, but not when the variable "*m*" is so replaced.

The converse of $\mathbb{C}$SS follows from earlier principles. For suppose not, and there is an *o* for which $\mathbb{C}om \wedge o \wr n$ but $\mathbb{C}mn$. Then by $\mathbb{C}$trans, $\mathbb{C}on$, which implies $o \odot n$ by $\mathbb{C}$refl and Df. $\odot$, and this in turn implies $\sim(o \wr n)$ by SepAx, in contradiction with the assumption $o \wr n$. So the strong supplementation principle can be strengthened to an equivalence.

The strong supplementation principle of standard mereology (together with the properties of parthood) entails the weak supplementation principle, the analogue of which is

$\mathbb{C}$WS $\quad \overline{\mathbb{C}}mn \supset \exists o\,(\overline{\mathbb{C}}on \wedge o \wr m)$.

This follows here since $\overline{\mathbb{C}}mn$ is just $\mathbb{C}mn$ and $\sim\mathbb{C}nm$ by Df. $\overline{\mathbb{C}}$, and the latter implies $\exists o\,(\mathbb{C}on \wedge o \wr m)$ by $\mathbb{C}$SS.

The analogy with mereology can be pursued further with the introduction of an axiom schema supporting the existence of micro- or quasi-sums. A direct analogue of schema MQ3 (Sect. 1.2) is not appropriate here, however, where there is no question of proving uniqueness. The quasi-operation would generate a microentity from microentities satisfying the condition $\varphi(m)$, and a microentity is not involved in meaningful identity statements. But the criterion of indistinguishability serves as a basis for establishing the indistinguishability of quasi- or microsums. We proceed by first establishing an alternative formulation of IndistAx:

T3 $\quad \forall m (o \wr m \equiv o \wr n) \supset m \approx n.$

From SepAx and Df. $\odot$ we have

$$\begin{aligned} \forall o (o \wr m \equiv o \wr n) &\equiv \forall o (\exists o' (\mathbb{C}o'o \wedge \mathbb{C}o'm) \equiv \exists o' (\mathbb{C}o'o \wedge \mathbb{C}o'n)) \\ &\equiv \forall o \forall o' (\mathbb{C}o'm \equiv \mathbb{C}o'n) \\ &\equiv \forall o' ((\mathbb{C}o'm \supset \mathbb{C}o'n) \wedge (\mathbb{C}o'n \supset \mathbb{C}o'm)) \\ &\equiv. \forall o' (\mathbb{C}o'm \supset \mathbb{C}o'n) \wedge \forall o' (\mathbb{C}o'n \supset \mathbb{C}o'm) \\ &\supset. (\mathbb{C}mm \supset \mathbb{C}mn) \wedge (\mathbb{C}nn \supset \mathbb{C}nm) \\ &\supset. \mathbb{C}mn \wedge \mathbb{C}nm \qquad \text{[Crefl]} \end{aligned}$$

which proves the theorem.

A microsum of φ-ers is now introduced as a microentity quasi-separate from all and only those things quasi-separate from each φ-er. With no identities amongst microentities, sums cannot be defined as in ordinary mereology (Sect. 1.2). Instead, the general microsummation operator $\Sigma m\, \varphi(m)$ is introduced as a primitive concept whose sense is determined by the axiom schema:

SumAx $\quad \exists m \varphi(m) \supset \forall m (m \wr \Sigma n\, \varphi(n) \equiv \forall n (\varphi(n) \supset m \wr n)).$

The same symbol, Σ, has been used as for classical mereological sums of quantities and the ensuing variable is taken to distinguish the cases.

The indistinguishability of microsums is rather cumbersome to formulate but easily proved on the basis of the criterion provided by T3. Abbreviating $\forall m (\varphi(m) \supset o \wr m)$ as $\psi(o)$, the theorem can be expressed as

T4 $\quad \forall n_1 \forall n_2 (\forall o (o \wr n_1 \equiv \psi(o)) \wedge \forall o (o \wr n_2 \equiv \psi(o)) \supset n_1 \approx n_2.$

Consider, then, any n_1, n_2 such that $\forall o (o \wr n_1 \equiv \psi(o)) \wedge \forall o (o \wr n_2 \equiv \psi(o))$

$$\begin{aligned} &\equiv \forall o ((o \wr n_1 \equiv \psi(o)) \wedge (o \wr n_2 \equiv \psi(o)) \\ &\supset \forall o (o \wr n_1 \equiv o \wr n_2), \end{aligned}$$

which implies $n_1 \approx n_2$ by T3.

Binary microsums, $n \cup o$, are defined by the condition $m \approx n \vee m \approx o$ of being indistinguishable from either n or o. The existence prerequisite is automatically fulfilled [$n \approx n \Rightarrow \exists m(m \approx n) \Rightarrow \exists m(m \approx n \vee m \approx o)$]. Clearly, $n \cup o \approx o \cup n$. Further,

T5 $\quad n \approx o \supset m \cup n \approx m \cup o.$

For given $n \approx o$, $m \cup n$ is $\Sigma m'(m' \approx m \vee m' \approx n)$ [by definition] $\approx \Sigma m'(m' \approx m \vee m' \approx o)$ [since $m' \approx o$ is equivalent with $m' \approx n$ by transitivity and symmetry of $\approx$] $\approx m \cup o$. Also,

T6 $\quad o \wr m \cup n \supset . \ o \wr m \wedge o \wr n.$

For by the SumAx,

$$\begin{aligned} o \wr m \cup n \equiv \ & \forall o' \, (o' \approx m \vee o' \approx n \, . \supset \, o \wr o') \\ \equiv . \ & \forall o' \, (o' \approx m \supset o \wr o') \ \wedge \ \forall o' \, (o' \approx n \supset o \wr o') \\ \supset . \ & (m \approx m \supset o \wr m) \ \wedge \ (n \approx n \supset o \wr n) \\ \supset . \ & o \wr m \ \wedge \ o \wr n \end{aligned}$$

(The analogous proof in standard mereology with identity instead of indistinguishability maintains the equivalence.)

A special case of a microsum, called the quasi- or microproduct of φ-ers and written $\Pi m \, \varphi(m)$, is the sum of microentities satisfying the condition of composing every φ-er, i.e. $\Sigma m \, \forall n \, (\varphi(n) \supset \mathbb{C}mn)$. The corresponding existence condition is, accordingly, $\exists m \, \forall n \, (\varphi(n) \supset \mathbb{C}mn)$ and SumAx becomes in this special case

$$\exists m \, \forall n \, (\varphi(n) \supset \mathbb{C}mn) \supset \forall m \, (m \wr \Pi n \, \varphi(n) \equiv \\ \forall n \, (\forall o \, (\varphi(o) \supset \mathbb{C}no) \supset m \wr n)).$$

Again, the same sign, Π, is used for the microproduct as for its mereological analogue, relying on the variable type to distinguish the cases. A binary microproduct, $n \cap o$, is defined by the condition of being indistinguishable from either n or o.

A quasi- or microdifference, $m - n$, between two microentities is what is left when n is taken away from m. This presupposes that there really is something left, i.e. $\exists o \, (\mathbb{C}om \wedge o \wr n)$, and a difference is a microsum of all these left-overs, $\Sigma o \, (\mathbb{C}om \wedge o \wr n)$. By the analogue of the strong supplementation principle, $\mathbb{C}$SS, which we saw could be strengthened to an equivalence, the presupposition is equivalent with $\sim \mathbb{C}mn$, and SumAx becomes in this special case

$$\sim \mathbb{C}oo' \supset \forall m \, (m \wr o - o' \equiv \forall n \, (\mathbb{C}no \wedge n \wr o' \, . \supset \, m \wr n)).$$

Note that $\sim \mathbb{C}mn$ allows that $m \odot n$, and where the more specific case of n properly compsing m is at issue the further prerequisite $\mathbb{C}nm$ must be specified.

What of the seemingly obvious *principle of reunification*

$$\overline{\mathbb{C}}nm \supset (m - n) \cup n \approx m,$$

where n properly composes m? With T3 in mind, it has to be shown that the indiscernibles here are quasi-separate from exactly the same things. Assume $\overline{\mathbb{C}}nm$, so that $m - n$ exists. What is quasi-separate from m has no composing microentity in common with m, and so by T1, nothing in common with anything composing m. In particular, it has nothing in common composing anything composing m and quasi-separate from n. So what is quasi-separate from m is quasi-separate from $m - n$ by SumAx as well as being quasi-separate from n in virtue of $\mathbb{C}nm$ and T1. But the conclusion doesn't follow in the absence of the converse of T6, which would seem to be unobjectionable and could be adopted as an axiom.

The converse,

$$\overline{\mathbb{C}}nm \supset \forall w\,(w \wr (m - n) \cup n \supset w \wr m),$$

is taken as axiomatic, and the principle of reunification is accepted.

A universe of microentities is a sum of everything satisfying a universal condition satisfied by all and only microentities. Being indistinguishable from itself won't do as the condition since it is also satisfied by macroentities. But a universe of microentities, U_m, can be defined as $\Sigma m\,\mathbb{C}mm$. (Present assumptions don't allow a meaningful formulation of a uniqueness claim.) A microcomplement, $-m$, of m is a difference with respect to a microuniverse, $U_m - m$, provided $\sim(m \approx U_m)$.

The principle of insensitivity to permutations mentioned at the beginning of the section can now be formulated, adding the proviso needed to guarantee the existence of a difference, $\sim\mathbb{C}mn$, which is tantamount to strengthening $\mathbb{C}nm$ to $\overline{\mathbb{C}}nm$:

$$n \approx o \;\wedge\; \overline{\mathbb{C}}nm \;\wedge\; \sim\mathbb{C}om \;.\!\supset\; (m - n) \cup o \approx m.$$

In view of T5, it follows that

$$n \approx o \;\supset\; (m - n) \cup o \approx (m - n) \cup n.$$

The reunification principle $\overline{\mathbb{C}}nm \supset (m - n) \cup n \approx n$ then gives us, allbeit somewhat clumsily, the principle of insensitivity to permutations.

7.4.1 Discussion

The quasi-mereology of indiscernible particles is, perhaps, too unlike mereology to be of interest. But what led to this investigation was the idea that parts at issue in distributivity properties that are a feature of the macroscopic behaviour of matter are themselves macroscopic entities. Microscopic constituents are not counterexamples because they are not parts of quantities. On this account, microentities are not parts because they don't respect the identity of parts which is the basis of mereology.

There is the question of what lies at the border between micro- and macroentities. Penrose gives the impression in the passage quoted above that the domain of microentities encompases ever larger compound microentities—the "etc." in "for the whole atoms, molecules, etc.". But at some point, in some way, the familiar rules of identity come into play. How this accommodation between the realms of the microscopic and the macroscopic comes about is an issue that lies beyond the scope of this book, however. At all events, no postulate is introduced to the effect that every quantity is a sum of mereological atoms, composed of microentities but with no proper parts. Still less are such mereological atoms identified with molecules or any kind of microentity.

Some understanding of this unhappy state of affairs might be gleaned from the standard approach to molecular structure in quantum chemistry. Electrons in molecules are treated as indistinguishable particles in the sense of quantum mechanics in modern theories of molecular structure. Nuclei are taken to form the framework—giving the molecule a shape—with reference to which the electronic structures are calculated. As such, nuclei are not treated as indistinguishable particles in molecular orbital theory, on the strength of what is variously referred to as the Born-Oppenheimer approximation or assumption. It is a contentious issue whether this can be justified in quantum-mechanical terms as an approximation or is more appropriately regarded as an independent assumption (Sutcliffe and Woolley 2012). The first accurate treatment of the isolated hydrogen molecule allowed permutation of the nuclei (James and Coolidge 1933), and Woolley (1988) points out that isolated molecules in gases at low pressure can be successfully treated (e.g. in mass spectrometry) without appeal to the Born-Oppenheimer assumption. The characteristic structure endowed by "clamped nuclei" commonly referred to as molecular shape is, it seems, a special feature of a molecule endowed by other molecules in close vicinity.

Chapter 8
Longish Processes

8.1 Occurrents

Three of Aristotle's elements, water, air and earth, leave some vestige in modern thought as kinds of matter: a particular phase of a compound (water), a solution of gases (air) and a very complex mixture of substances. It remains to come to terms with the fourth Aristotelian element, fire. As with the other Aristotelian elements, there is something in the idea. But it is not a kind of quantity of matter. It is a process.

Processes have temporal (and other) parts and do not endure over time like continuants. They are what Johnson called occurrents. As mentioned at the beginning of Chap. 1, they figure in reductionist theses, some philosophers suggesting that they are reducible to continuants and others suggesting that continuants are reducible to them. Neither thesis is advanced here, and processes are included in the basic ontology as a separate domain over which range a distinct style of variables, $e, f, g, \ldots$ of the many-sorted regimentation. Whether some deep argument establishing the reduction of one category of entity to another may yet be forthcoming cannot, of course, be ruled out. In the present case, the interesting candidate seems to be the reduction of thermodynamics, which provides much of the motivation here for the introduction of processes, to statistical mechanics. But even if macrotheory were reducible to microtheory, that would not eliminate the ontological claims of existence of macroentities nor show that their properties were somehow not features of the world. As things stand, however, in the absence of any such argument,[1] the presumption is that processes constitute an autonomous category.

[1]Even friends of reduction feel bound to admit "the jury is still out concerning the reduction of thermodynamics to SM [statistical mechanics]" (Callender 1999, p. 352).

P. Needham, *Macroscopic Metaphysics*, Synthese Library 390,
https://doi.org/10.1007/978-3-319-70999-4_8

Philosophers writing about occurrents often speak of events and their examples are often human actions or occurrences suffered by humans, with descriptions couched in terms that focus on their human interest. Speaking of processes is a more natural way of describing what, from the macroscopic perspective, are continuously developing progressions and examples involving human agents are avoided in order not to obscure the physical aspects of processes. This in no way prejudges the issue of an identity thesis along the lines Davidson has proposed, taking actions to be processes as understood here, but that is not the present concern. Nor should it obscure the fact that one and the same general thesis about the nature of occurrents may be at issue whether formulated in terms of events or by speaking of processes. One such issue is the reductionist strategy that events are but changes in continuants. This is not a deep theoretical argument like that alluded to in the previous paragraph, but a way of essentially trivialising the notion of an occurrent. As an analysis it fails because, whilst a process may determine a change, a change does not determine a specific process. This conception is taken up and the criticism elaborated in the next section. It is followed in Sect. 8.3 by an account of how the notion of a process was encapsulated right from the first stages of the development of thermodynamics as it dispensed with caloric in dealing with the distinctions of warmth and heating. A process cannot, on this account, be adequately characterised solely by conditions at its beginning and end, which is how those who define occurrents as changes would have it. Some indication is given of how the conception is developed in the extension of the theory to irreversible thermodynamics, which is applicable under conditions not too far from equilibrium, with the notion of entropy production.

Since the action of the entropy-producing forces driving processes is naturally described in causal terms, the relation of processes to causation is taken up in Sect. 8.4. There is no claim to give an analysis of causation, however. The aim of the account of processes presented here is therefore to be sharply distinguished from that of another conception of processes initially developed by Reichenbach and Salmon from a very different point of departure, which is widely held to have failed as an analysis of the concept of causation (Psillos 2002, pp. 110–27). Reichenbach and Salmon's concern with world lines is not an issue here, where the mere endurance of a continuant is not regarded as a process. Nor is the mere conservation of energy, incorporated into latter-day modifications of such theories, regarded as sufficient to merit speaking of a process. The mere conservation of energy in accordance with the first law of thermodynamics is compatible with the persistence of temperature gradients, the uneven distribution of a substance and the persistence of iron in the presence of oxygen[2] despite iron oxide having proportionately less energy. Only the second law provides anything like the basis for an explanation of why processes of heating, diffusion and chemical transformation occur.

[2]To take the example Atkins (1994, pp. 111–4) uses to illustrate the way in which entropy drives chemical reactions in general.

In his review of causal process theories, Dowe (2009, p. 223) takes up the claim he and Salmon made about their Conserved Quantity theory providing an empirical analysis ("it concerns an objective feature of the actual world ... draw[ing] its primary justification from our best scientific theories"), which they contrast with a conceptual analysis, tied to uneducated folk intuition and serving a different aim. A conceptual analysis of change and event found in the philosophical literature is challenged in the following pages by an empirically justified conception of processes based on ideas developed by the proponents of thermodynamics. This calls for broad conceptual strokes rather than quantitative detail, so does not raise detailed questions of empirical justification. But the conception avoids the fixation on mechanical collisions typified by the kinds of process theory Salmon has inspired, which involves a restriction on the claim of empirical adequacy that Dowe radically underestimates when he goes on to say that "the conserved quantity theory ... requires commitment to a fairly thoroughgoing reductionism" (2009, p. 224). It is not just economics and psychology, as he seems to suggest, that would have to be reduced, but also vast tracts of natural science. The theory doesn't touch chemical reactions (and therefore subjects such as biology or mineralogy), chemical change resulting from the deformation of materials (Bayly 1992), phase change and much else. In fact, nearly all of the macroscopic realm investigated by the physical sciences would have to fall to the as yet unproven reductionist thesis if the claim of empirical justification were to be borne out.

Davidson's strategy of introducing processes as variables of quantification about which several things can be said in defence of the relational view of causation as a dyadic relation between events offers something more useful to the present project. The general strategy itself is not tied to the relational view of causation, from which it is severed in Sect. 8.4, freeing it for use as a vehicle for describing how processes are related to continuants, space and time in Sect. 8.5 without the Davidsonian trappings. The mereological structure of parts—temporal and otherwise—of processes is taken up in Sect. 8.6, after which examples illustrating the relational character of processes are discussed in Sect. 8.7 both for their intrinsic interest and by way of putting the final nail in the coffin of the idea underlying some philosophical treatments of events that they involve a single continuant.

8.2 Change

Philosophers seeking to define occurrents as changes find fault with Russell's view that a change can be fully characterised as "the difference, in respect of truth and falsehood, between a proposition concerning an entity and a time T and a proposition concerning the same entity and another time T′, provided that the two propositions differ only by the fact that T occurs in one where T′ occurs in the other" (1972, p. 469) because it fails to distinguish "real" changes from "Cambridge" changes. The leading idea in Lombard (1986), for example, is that real changes involve non-relational changes in objects, whereas the proposition in Russell's definition may well involve other entities than the one mentioned. Cleland (1991) thinks it a mistake to invariably ascribe change to physical objects, and considers it a problem for Kim's (1976) analysis of events as property exemplifications that

> some events can be identified without reference to any physical object (e.g. that shriek, this flash, that desire), which suggests that some events may not be individuated in terms of physical objects. (1991, p. 230)

Her leading idea is a refinement of Bennett's (1988) theory of events as tropes (instantiations of properties), which is itself a refinement of Kim's theory. An event is a pair of exemplifications of distinct states at distinct times by a kind of trope she calls a concrete phase (a particularised determinable, of which temperature is taken to be an example).

Lombard argues that events are non-relational Russellian changes on the basis of examples such as Socrates dying, which entails that his wife, Xanthippe, becomes a widow, or that while Theaetetus grows, Socrates, who was taller than Theaetetus becomes shorter. Unlike Xanthippe's becoming a widow and Socrates' becoming shorter than Theaetetus, which involve changes in the relation of Xanthippe to Socrates, and of Socrates to Theaetetus, Socrates' dying and Theaetetus' growing are said to be genuine changes—things that happen or occur—because they are non-relational. But it seems that Lombard is too hasty in generalising from these cases. Many processes are naturally regarded as relational. There are processes of mutual attraction and repulsion of electrically charged and magnetised objects, and of Newtonian gravitational attraction involving massive bodies such as the earth and the moon, the sun and the planets, and so forth. The collision of two billiard balls involves the impressing on and thrusting away of one by the other. There is heating of a colder body by a warmer body and the diffusing of a quantity of a substance through another, as well as the chemical reaction of several substances yielding others. In each of these cases, some feature of the states of the bodies involved (electric charge, magnetic pole, mass, uniform velocity, temperature, concentration, chemical potential) gives rise, in the circumstances, to the processes.

Despite the prevalence of such examples, philosophers have been reluctant to take them at face value and adopt a relational view of processes. Traditionally, they have imposed a different view on examples of the kind just mentioned according to which there are several changings, and so events, each involving a change in a single body rather than relations between several bodies. It is said, for example, that there are changings in position of the billiard balls, or of the temperatures of the warmer and the colder bodies in the heating case, and the relational aspect is a matter of these changings standing to one another as cause and effect. The general strategy of construing events in whatever way is deemed necessary in order to save the relational view of causation as a dyadic relation between events will not go unquestioned here. But suffice it for the moment to mention two examples indicating some limitations on the workability of this idea. First, in the paradigm example of one billiard ball striking another, it is unclear how the approach is to be generalised to cover three-body collisions. No change in any single body can be identified with the cause on any view of the matter.

In the second case, the temperature difference between the bodies involved in a heating might be maintained by thermostatically controlling each body, and yet a heating continues to take place. Duhem describes such a case:

> Let us imagine a metal bar whose surfaces are surrounded by a non-conducting substance which doesn't allow any exchange of heat between the bar's surface and the external environment. One end of the bar is at a base temperature ϑ, the other is submitted to the action of a source of heat with a temperature ϑ', higher than ϑ. After a certain time, a permanent regime is established. The temperature of each point in the bar and the state of each element of volume of the bar then remain invariable. If, therefore, starting from the moment that the permanent regime is established, the bar is observed for a certain time, its internal energy will not vary, the external forces acting on it will do no work, its kinetic energy will remain equal to 0. At the same time, the bar will have absorbed heat at its hot end and released heat at its cold end. This is the simple observation that Clausius has generalised in such a way as to give ... [his] fundamental postulate (Duhem 1887, p. 126).

So a change in temperature is by no means essential to the occurrence of the heating process. Thermodynamics suggests that change is involved in the environment of the bar. But this sounds like the sort of Cambridge change that was to be avoided. In any case, it is not at all obvious how this insight can be harnessed in defining the heating process as a change in an object. Which object would that be? Neither the bar nor its parts, since their temperatures are not changed, although the bar and its parts are what, if anything, is heated.

The view of occurrences as non-relational changes faces many apparent counterexamples, and is not at all adequately motivated by the usual scare stories about Cambridge changes. I turn to consider Cleland's view of occurrents in more detail.

The counterexamples Cleland adduces against the thesis that events involve physical objects do not clearly bear out her claim. It may not be apparent to an observer what it is that shrieks or flashes, but they might nevertheless be compression waves or ionisation effects imparted to the air, or dispersion of light by smoke. And it begs a large question to assume that desires are not features of material. Still, there might be processes which can't be interpreted as going on in material. Radiation phenomena, for example, might simply be borne by regions of space at a time.[3] Even so, there are problems with her definition of occurrents.

The properties whose gain and loss determine what counts as a change Lombard qualifies as static. Cleland calls them states, and defines a change in terms of an initial and final state. She claims to be following standard scientific practice by appealing to the notion of states and their relation to changes as characterised in science. Simons (2003, p. 369) agrees that "her account accords closely with the way physicists specify the states of physical systems in terms of the values of variable quantities", which he counts as a point in its favour (without committing himself

---

[3]"... the empirical equations of state of ... an electromagnetic cavity are the 'Stefan-Boltzmann Law' $U = bVT^4$ and $P = U/3V$ where b is a particular constant .... It will be noted that these empirical equations of state are functions of U and V, but not of N. This observation calls our attention to the fact that in the 'empty' cavity there exist no conserved particles to be counted by a parameter N" (Callen 1985, pp. 78–9). Cf. what Lieb and Yngvason (1999, pp. 20–1) say about the thermodynamic treatment of electromagnetic phenomena.

to it). At one point, Cleland says that "[C]hange is commonly described in science [by being] represented in a state space ... as a 'trajectory' (time-ordered curve) connecting different states" (1991, p. 242). This comment reflects an important insight, but it is not captured by her definition. A concrete change (which an event is subsequently defined to be), she says,

> is a pair $\{x, y\}$ such that x is the exemplification of a state s by a concrete phase CP [determinable of which the state is a determinate feature] at a time t and y is the exemplification of a state s' by concrete phase CP$'$ at a time t', where (i) t is earlier than t'; (ii) CP is the same concrete phase as CP$'$, and (iii) s is not the same state as s'. (Cleland 1991, p. 238)

Condition (i) precludes instantaneous changes, which I take to be a condition of adequacy of any acceptable notion of process. But this is not enough to capture the idea of a "trajectory". Condition (iii) excludes the important case of cyclic transformations, which follow a closed trajectory, i.e. one returning a system to its initial state. A good indication that the import of processes taking time is not properly captured is that the definition fails to allow for the distinction between processes where the initial state is merely recovered and reversible processes (discussed further in the next section), which specifically involve retracing the trajectory in recovering the initial state. Carnot cycles are processes combining both these features, i.e. processes following closed, reversible trajectories. Cyclic and reversible processes are special cases of processes with duration which are represented by time-ordered curves connecting specific states.

In general, two states might be connected by any number of trajectories, each corresponding to a possible happening, no one of which can therefore be identified merely by reference to its endpoints. A mountaineer might be at a certain height at one time and another at a later time, but this tells us little about the precise route traversed during the intervening period and the particular activities involved in it, notably the smaller ascents and descents which sum to give the total change in height. It doesn't even distinguish the honest toils of a mountaineer from the cheat who ascends from the lower height in a balloon and later descends to the higher point. Of course, a particular change of state from some given state of a body to another is the result of some actual process, which can therefore be correctly described as the process leading from such-and-such a state at a particular time to such-and-such a state at another time (the times must be specified since states are repeatable). But whatever matters of reference this strategy addresses, it doesn't yield general features specifying a process as being of some kind or other which would facilitate scientific theorising about processes, contrary to Simons' assessment. Moreover, it cannot be used to specify a possible process. Given the fact that formal treatments of thermodynamics deal with a relation of adiabatic accessibility on a set of all possible states, this problem is not lightly dismissed.

A description in terms of process kinds must involve a predication of the entire period and can't in general be characterised by simply specifying the states obtaining at its endpoints. This is the first problem with Cleland's definition. The second is that, although she happens to have chosen an example of a state determinable when illustrating her definition with temperature, being a state determinable imposes special conditions. A general metaphysical theory which simply treats them as any old determinable (what Cleland would call a phase), such as heat and work, doesn't fit the bill.

I don't want to go into the general metaphysical theory of tropes on which Cleland builds her conception of processes as changes. But there is one aspect of the motivation she gives for this approach to the problem which is relevant to this second line of criticism. Cleland says

> Remember that old puzzle about change: in order for something to change there must be a sense in which it remains the same (otherwise it simply ceases to exist) and a sense in which it becomes different (changes). There is a sense in which a concrete phase can be said to remain the same while becoming different. It can be said to remain the same in the sense that the instance of the determinable property which constitutes it does not go into and out of existence during the time in which the instances of the associated determinate properties (states) are going into and out of existence, e.g., a liquid whose temperature is changing does not, at any time during the change, cease to have temperature .... it can also be said to become different in the sense that it takes on different determinate values (e.g. of temperature) during the change. More specifically, if one takes determinate properties (states) to be contingent properties of instances of the determinable they fall under (of concrete phases), then it seems that one can make good metaphysical sense of the Aristotelian claim that the proper subjects of change are conditions, viz., concrete phases, as opposed to physical objects. (Cleland 1991, pp. 237–8)

Temperature is an intensive property, its application requiring that all spatial parts of whatever bears a temperature have the same temperature as the whole. In order to have a temperature a body must therefore be at thermal equilibrium. If a body is to have a temperature at any time during a change this condition must be maintained during the change. The conditions of so-called quasi-static change when this constraint is satisfied represent an ideal which can only be approached in practice, and ordinary, everyday changes in temperature are not quasi-static. This means they are not even representable by paths through state space, the intermediate conditions not being equilibrium states. Again, there are any number of irreversible processes with the same initial and the same final states, and distinguishing any one of them from the rest must take account of what happens throughout the entire period. Experimentally, raising the temperature of a liquid would require waiting at the end of the process for the liquid to settle down to equilibrium and acquire a definite temperature. So it is not clear that it is true to say of any proper subintervals of this period that it is "a liquid whose temperature is changing" when it doesn't have a temperature at the start and end of the subinterval in question. The process nevertheless involves something which "remains the same", but the only serious candidate in the offing is the quantity of liquid, i.e. the physical continuant.

What Lombard offers is no improvement on Cleland's notion of states. He distinguishes what he calls static properties as those whose being instantiated at an instant does not imply the object in question is changing. Instantaneous dynamic properties, by contrast, entail the existence of an interval including the instant in question during which the object has and then lacks some static property. Acceleration during a period which is part of a longer period of acceleration—for example, the moon's revolving around the earth all day yesterday—does not involve a dynamic property by this criterion since there is no static property acquired and then lost.

I have not pursued how the issue might be affected by any specific account of the general import of predicates, such as trope theories. But I will mention that in a recent review of attempts at drawing a useful distinction between intrinsic and extrinsic properties, Weatherson (2006, p. 6) observes that distinguishing a suitable notion of non-relational property falls back on a theory of identity conditions for states and events. The policy here will be to introduce processes outright, rather than attempting to define processes as some kind of change, and seek a better characterisation of their central features than is accomplished by these strategies of definition, again without controversial assumptions about the import of predicates.

8.3 Distinguishing Processes from States

Heating is the paradigm case of a kind of process. We saw in Sect. 6.2 something of how the idea of heat as a substance was introduced in the latter part of the eighteenth century, described by Lavoisier as the "element of heat or fire" (1965, p. 175) for which he introduced the name "caloric". Thermodynamics developed in the mid-nineteenth century by incorporating the insight that the defining condition of caloric, that heat is conserved, is not true. The first law of thermodynamics introduced the concept of energy as a conserved magnitude, but instead of regarding energy as a kind of substance transferred from one body to another, treated it as measurable property characterising the *state* of a body (the retention of textbook talk of transferring energy notwithstanding). Heating was treated on a par with mechanical working, as a feature of the *processes* by which a body is changed from one state to another. A given state may result from any number of different processes, and the given state of a body contains no trace of which particular process brought it to that state. In fact, a body at equilibrium (and so in a definite state) contains no trace of its history, not even how long it has been in the given state. So it is already becoming clear that an end state is in no way characteristic of a process. The original distinction that first motivated the caloric theory, the need to clearly distinguish the notions of warmth and heat, remains in thermodynamics but is incorporated into a general theory in a different way—one which remains the basis of our understanding of temperature and heating today. These developments are reviewed in this section with a view to showing how ontological commitment to processes arose as a result of distinguishing the way processes and states are described.

The early designers of thermometers were aware that these instruments measure the degree of warmth of a body without providing any way of answering the further question of whether or not the same amount of heat is required to raise a given body through a given interval of temperature, say one degree, in different regions of the temperature scale, or whether the same amount of heat is required to raise different bodies through the same degrees, say from 0 to 10 °C. Such questions were dimly understood at first. But the introduction of the concept of specific heat, and above all Joseph Black's discovery of latent heat, established beyond doubt that a distinction of some sort is needed between the degree of warmth of a body, recorded by a thermometer, and what was described as the transfer of heat to a body. This latter idiom persists in modern textbooks, although usually accompanied by a word of caution about the unwanted suggestion of the movement of some calorific substance from one body to another or the conservation of some feature of the motion of microscopic parts. The downfall of the caloric theory was not so much due to the ethereal qualities of the imponderable substance, as to a failing it shared with the competing theory at the beginning of the 19th century according to which heat is a movement of the constituent particles of a body. As Duhem put it,

> if ... physicists were divided regarding the nature of heat, they were unanimous in recognising that the heat absorbed by a body during a transformation depends only on the state from which the body departed and that to which it arrived. They were unanimous in proclaiming that during the traversing of a closed cycle [in which] a body [returns to the same state, it] releases as much heat as it absorbs. (Duhem 1895, p. 403)

But this proved inadequate; "it is necessary", Duhem continues, "to recount ... the history of the body from the start of the change until the end", and

> It is no longer possible to speak simply of a body, taken in a given state, containing a determinate quantity of heat; for, if it is brought back to this state after having undergone a series of changes, it will in general have absorbed more heat than it has released, or released more heat than it has absorbed. It is still possible to speak of the quantity of heat absorbed or released by a body while it is undergoing a determinate transformation. But it is not any longer possible to speak of the quantity of heat enclosed in a body taken in a given state (Duhem 1895, p. 404).

There is a kind of process, appropriately called heating, which is not a state or a function of the state of a body like its degree of warmth. But there is no such thing as the state of a body known as its heat or the heat content of a body, and thus no such thing as the transfer of heat from one body to another.

The first law of thermodynamics overturned theories ascribing bodies an amount of heat on the basis of a general distinction between features of state, which won't do to characterise processes, and features of processes, which cannot be used to characterise states. The mathematical form of the argument establishes that the sum of the infinitesimal amount of heating and the amount of working in an infinitesimal process has the properties of an exact differential—whose line integral over a closed path is zero, or equivalently, whose integral along a path depends only on the end-points of the path—from which it follows that there exists a function, called the energy, characterising a state. This is what the energy conservation thesis amounts to: there are reproducible conditions of bodies called states, and returning to a given state means returning to the same amount of energy. The energy difference between two states therefore imposes a constraint on any process in which a body changes from one of these states to the other: the sum of the amounts of heating and of mechanical working involved equals this difference in energy between the states. But these amounts of heating and of mechanical working may vary from one such process to another, increases in the one being compensated by decreases in the other in order that their sum always equals this difference in energy between the states. Accordingly, the change in energy (or any other feature of state) is not sufficient to distinguish any one of an infinite number of different possible processes carrying the body between these states from any other, and the state of a body bears no trace of how it got into that state.[4] Equally, characteristic properties of processes such as the amount of heating involved won't do to characterise a change of state because they say nothing of how much of some other feature of the process such as mechanical working is necessary to offset the heating and yield a given change of energy. This is the interpretation of the mathematical feature that functions tracking the changes in features, such as in the amount of heating, or the amount of working, characteristic of processes carrying a body from one state to another are inexact differentials, whose integrals between two limits (states) are dependent on the path taken.

Thermodynamic data are exclusively concerned with equilibrium states, and used to lay down necessary conditions for the spontaneous occurrence of processes, which cannot transform material to a less stable condition. But although primarily concerned with states of bodies, it is clear from the foregoing that classical thermodynamics prepares the ground for a more explicit treatment of processes by the way it distinguishes properties of state from those of processes. Without full-blown recognition of processes as things taking place in time, however, it raises a ticklish problem in connection with its core notion of a reversible process, which is central to classical arguments based on Carnot cycles and the establishment of properties of state. The notion was in general usage before being carefully defined, apparently for the first time by Duhem (1887, pp. 132–4; 1893a, pp. 305–9). Rechel (1947, p. 301) drew attention to the nice point involved when he bemoaned the fact that the standard English term "reversible process" "contains a contradiction within itself... which text book writers are prone to ignore", although he exonerates Duhem from the charge.

Duhem notes, by way of a preliminary to his definition, that

> When a system is at equilibrium in a certain state it persists indefinitely in that state, and is not transformed. It therefore seems not possible to speak without contradiction of a real change constituted by a sequence of equilibrium states. But it is, in fact, possible to give these words a logical meaning. (Duhem 1893a, p. 302)

The process of allowing a gas to expand by reducing the pressure on a piston to a value below that of the internal pressure of the gas is irreversible, giving rise to non-uniformities depriving the gas as a whole of well-defined pressure and temperature. By successively reducing the difference between the external pressure and the opposing internal forces, the gas expands more slowly and less chaotically. The amount of work done by the gas increases to a limit in the case where the external and internal pressures are equal. But in this limiting case, the gas is at equilibrium,

---

[4] On this basis, the distinction between liquid and gas phases is argued to be unclear, at least as a characterisation of the state of a body, and no finer description of state than "fluid" is really justified. Isothermal compression of a gas below its critical point leads to drops of liquid appearing as the gas condenses, and eventually the last trace of gas disappears. But the same highly compressed state may be reached via an alternative route (a different kind of process)—increasing the temperature to above the critical point at constant volume, then cooling at constant pressure, followed by cooling at constant volume—in which no condensation occurs (Castellan 1964, p. 36).

and will remain in this state until such time as the equilibrium is disturbed. Real processes only occur as a result of an imbalance of forces between a system and its environment. Duhem suggested that in the limit, now called a quasi-static process after Carathéodory (1909 [1976] p. 238), in which the imbalance is successively reduced, each step is an equilibrium state. A reversible process is one in which the quasi-static limit of successively reducing the imbalance of internal and external forces in taking the system from the final state to the initial state is the same set of equilibrium states, but traversed in reverse order. Accordingly, a quasi-static process is not necessarily reversible, as Duhem illustrates: "cycles of *magnetic hysteresis*, for example, are irreversible changes although they are continuous sequences of equilibrium states" (1893a, p. 307).

Duhem speaks of a reversible change (*modification réversible*) in the 1893a article, and suggests that, although not a real process, a sequence of equilibrium states can be regarded as a virtual change, which doesn't occur in time with the independent variables on which the state of the system depends being functions of time and having determinate rates of change. But this reversible "change" may be cyclic, and is thus not a change as Russell defined it. The standard English term "reversible process" is applied because any continuous path connecting one state with another is called a process, and implies a special kind of process. But the contradiction Rechel points to arises because of an equivocation between this use of the term "process" and the ordinary usage in which it cannot be applied to a sequence of equilibrium states because no system would move spontaneously from the one to another. The term "process" is understood here in this latter sense, to apply to something that can actually occur in time. In the absence of any suitable alternative to "change" and "process", I suggest that the terms "reversible process"' and "quasi-static process" be understood like the term "broken glass": just as a broken glass is not a glass, so a reversible process is not a process. Accordingly, since the state space of thermodynamics is a space of equilibrium states, a trajectory may be a quasi-static process, but it is not a process.

This calls for some qualification on what was said about trajectories in the previous section. It was argued there that Cleland's definition didn't realise one of its declared aims of representing a change as a trajectory in state space. Now we have seen that tracking the development in time throughout the course of a real process cannot be done by following a path in state space.[5] It is assumed in classical thermodynamics that any change in state of a body resulting from a real process can be reproduced by a reversible process for the purposes of calculating properties of state, but this is, as Duhem puts it, a virtual process which doesn't reproduce the process itself. The emphasis on equilibrium, which came with Gibbs's famous paper from 1876–8, was not always how the creators of thermodynamics thought

[5]The fact that (real) processes are not quasi-static, and so don't trace a path in the space of equilibrium states, doesn't redeem the failure of the attempt to define occurrents as changes. Changes still don't determine processes. A non-equilibriium condition of a body may be the result of any number of irreversible processes, as may an equilibrium state, which may also result from any number of quasi-static processes.

of their subject. Although entropy is only defined in equilibrium thermodynamics for systems at equilibrium, when Clausius first introduced the term he considered it to concern irreversible processes. The Clausius inequality, that entropy changes of closed systems are always greater than or equal to $\int đQ/T$, he put in the form of an equality in which what he called the uncompensated transformation (he coined the term "entropy" from the Greek word for transformation) in a change from one state to another is the entropy, S, of the final state less the entropy, S_0, of the initial state less $\int đQ/T$. The latter term, $\int đQ/T$, is the entropy due to heating of or by the environment, which equals the difference between initial and final entropies in reversible processes but falls short of this in real processes. The Clausius inequality, $\int đQ/T \geq 0$, implies that the shortfall, or uncompensated transformation, is always positive for irreversible processes (Duhem 1887, p. 146). But he developed no theory for determining the magnitude of the uncompensated transformation associated with any irreversible processes. Duhem seems to be the first to have explicitly developed accounts of the entropy produced by processes involving heat conductivity and viscosity (Miller 1960; Brouzeng 1991; Bordoni 2012). Since then, a theory of irreversible or nonequilibrium thermodynamics has been developed which relates the change of entropy to variables that characterise irreversible processes, such as the rates of chemical reactions.

Nonequilibrium thermodynamics, which we have already briefly encountered in Sect. 3.5, builds on the ground prepared by equilibrium thermodynamics by treating intensive thermodynamic properties and the corresponding densities for the extensive properties such as entropy as varying with position and time on the assumption that local equilibrium always reigns at each point and varies smoothly from one to another. The assumptions of irreversible thermodynamics are not reasonable under conditions of turbulence, but are applicable under conditions not too far from equilibrium which are realised in many real situations. The theory deals with rates of change (in time) by considering the entropy change as the sum of the entropy of a reversible change taking the system at issue from the initial to the final state in question and the entropy created internally within the system, i.e. Clausius' uncompensated transformation. It strengthens the statement of the second law in equilibrium thermodynamics to encompass the parts of a system: not only is the total internal entropy production nonzero, but so too is that of each part (Kondepudi and Prigogine 1998, p. 90). It provides a general law giving the entropy production, σ—the rate of change of the entropy density (the internal entropy per unit volume)—as a sum of terms involving the product of a generalised force or affinity and a corresponding flow (Callen 1985, p. 308)[6]:

$$\sigma = \Sigma J_i F_i.$$

[6]Differentiating the entropy, $S(X_0, X_1, \ldots)$, expressed as a function of the extensive variables $X_0, X_1, \ldots$, which vary with time, we have $dS/dt = \Sigma_k \partial S/\partial X_k \, dX_k/dt$, where the dX_k/dt are the fluxes of X_k and the entropy-representation intensive magnitudes $\partial S/\partial X_k$ are the corresponding generalised forces or affinities. Omitting the volume from the list $X_0, X_1, \ldots$, this gives the rate of change of the entropy density, s, in terms of the extensive magnitude densities, x_k, as $\sigma = ds/dt = \Sigma_k \partial s/\partial x_k \, dx_k/dt$.

A flow, J_i, is a time rate of change which might be of the movement of matter or electricity, or of "heat" (i.e., a rate of heating), or rate of a chemical reaction converting some kinds of matter into others. The generalised forces, F_i, usually take the form of gradients of thermodynamic state variables such as the chemical potential, the electrical potential and temperature (usually expressed as a gradient of the inverse of the temperature, $1/T$). These gradients express in analytic terms the way the respective thermodynamic features vary over space. In the case of chemical reactions, the flow is the velocity of the reaction and the force is given by the chemical affinity of the reaction, which is not a gradient over space but defined essentially as the difference between the sum of the chemical potentials of the reactants and that of the products. For a reaction

$$a_1A_1 + \cdots + a_mA_m \rightarrow b_{m+1}B_{m+1} + \cdots + b_nB_n,$$

the chemical affinity or force is $-\Sigma\nu_k\mu_k$, where ν_k is defined in terms of the stoichiometric coefficients as $-a_k$ for a reactant, A_k, or $+b_k$ for a product B_k, and μ_k is the chemical potential of the kth substance.

Understanding each distinct flow, J_i, to represent the rate of occurrence of a distinct process, several processes may well occur in the same region during the same time. In that case, not only is each flow, J_i, linearly related to its conjugate force, F_i, but is also linearly related to all the other forces determining the entropy production, σ, although each flow depends most strongly on its "own" force. Thus,

$$J_i = \Sigma_j L_{ij} F_j,$$

where L_{ij} is the general linear coefficient. Many of these cross-effects have been known as isolated phenomena since the early nineteenth century, but were first given a unified treatment when the theory was formulated in the 1930s on the basis of a certain kind of symmetry expressed by the Onsager relations.[7] An example of a cross-effect is the Seebeck effect, discovered in 1821, in which heating, driven by its own force of a temperature gradient, gives rise to an electric current. Thus, if the junctions at the ends of a copper wire inserted into a circuit otherwise comprising aluminium wire are maintained at different temperatures by heat reservoirs, a current flows under an electromotive force, generally of the order of 10^{-5} volts per Kelvin of temperature difference. Conversely, in the Peltier effect, discovered in 1834, the same circuit is used except that the potentiometer measuring the electric potential difference for the Seebeck effect is replaced by a battery. The two junctions are now maintained at the same temperature, and a current is passed through the system. Here, the current drives a heat flow from one junction to the other, generally of the order of 10^{-5} Joule per second per amp. Thus, although often the source of heating, a temperature gradient is not the only cause of heating.

[7] According to which, if the J_i and X_i in the expression for σ are independent, the linear coefficients satisfy $L_{ij} = L_{ji}$, for all i and j.

8.4 Causings

Writers expounding the principles of irreversible thermodynamics fall naturally into the causal idiom. In their introduction to irreversible thermodynamics, for example, Kondepudi and Prigogine (1998) say that "a concentration difference between adjacent parts of a system is the thermodynamic force that causes the flow of matter" (p. 88) and "a nonvanishing gradient of $(1/T)$ causes the flow of heat" (p. 351). In general the flows are caused by the conjunction of all the circumstances on which they depend. "For example, in thermoelectricity", as Miller (1960, p. 17) puts it, "the flow of current is caused by the temperature gradient as well as the usual electric potential gradient". These claims are not accommodated on the relational view of causation since a state and not an event or process is said to cause the process.

This challenge to the view that causation is always to be construed as a dyadic relation between events won't impress anyone indifferent to the distinction between process and state, or who understands the distinction on the basis of certain criteria proposed in the philosophical literature which are not up to the task. Davidson's (1980, pp. 105–22) adverbial modification argument, for example, supposedly motivating ontological commitment to events, works just as well on sitting still, moving with a uniform velocity in a straight line, rotating uniformly about an axis, remaining totally immobile, lying in the morgue, etc., which would ordinarily be regarded as states, as it does on his example of an event, Jones' buttering the toast. Again, sentences can be nominalised in two ways. The sentence "Quisling betrays Norway" becomes what Bennett (1988, p. 5) calls the perfect nominal, "Quisling's betrayal of Norway", and what he calls the imperfect nominal, "Quisling's betraying Norway". It has been suggested that whereas the imperfect nominals stand for propositions, the perfect nominals refer to processes. But perfect nominals include expressions such as "the body's uniform motion in a straight line" and "the body's warmth", which are treated by the appropriate scientific theories as states. Whatever the distinction drawn by linguistic features such as these may amount to, there is no reason to think they capture the substantial distinction marked by scientific criteria.

A diffusion process, involving the flow of matter of some kind of substance k through another substance, is driven by the negative gradient (across space) of the chemical potential of substance k. If the process is allowed to go to equilibrium, the differences in chemical potential at different places are ironed out, in which case the gradient diminishes and the diffusion eventually peters out. But it is not the diminishing of the gradient that causes the diffusion at any moment. On the contrary, the gradient diminishes because of the diffusion. Similarly, the gradient of the inverse of the temperature is a measure of the increase in $1/T$ (decrease in T) at a point in each spatial direction (a vector). In Duhem's example described in Sect. 8.2 of the bar whose ends are maintained at constant temperature, the gradient is held constant and causes the heating to continue indefinitely. Usually, heating caused by a temperature gradient reduces the gradient until it disappears and heating ceases. Where the temperature gradient does vary with time, the gradient of $(1/T)$, and not the rate of change of the gradient of $(1/T)$ over time, remains the cause of the heating. What causes a process—whose rate of progress is measured by a flow—is not a change, but a state of a body, even when this is changing.

When left to proceed to equilibrium, processes eventually peter out and establish a stationary equilibrium state. Prior to the establishment of equilibrium, a fleeting state of reduced temperature gradient, partial combustion of the materials involved in a chemical reaction, and so forth, results. Processes are themselves causings—the continuous transforming of a condition in a body to another. Such conditions are idealised as states of the body, but are more realistically treated as patterns of states of small parts of the body, continuously varying over space, which adjust as the process proceeds unhindered until a uniform state of equilibrium under the prevailing constraints is attained. This is at odds with Davidson's view. He has argued that "it must in general be a mistake to suppose that whenever an event is caused there must be something called a causing". As he puts it, "[i]f such a third event were required to relate the original cause and effect, two more events would presumably be needed to relate the original cause and effect with the causing". If not vicious, he thinks this is at least uncomfortable with its unnecessary multiplication of entities, and concludes that " 'caused' relates events, as do the words 'before' and 'after'; it does not introduce an event itself" (Davidson 2004, pp. 102–3). But the argument is not, as he seems to realise, quite so cut and dried. The multiplication of entities is engendered by the view that causal scenarios should be uniformly analysed on the model of two events standing in a dyadic causal relation, which requires a corresponding reduction of causings. Davidson's argument merely shows that an adherent of this model, like himself, will have difficulty in accepting causings. It provides no motivation for the model itself, and is not the independent argument that would be needed to dissuade the friend of causings with no allegiance to the relational theory of causation.

Nor is the argument bolstered by other Davidsonian strategies. LePore has sought to combine Davidson's intuitions about causation with the adverbial modification argument, which he takes to be among the "strongest arguments for the existence of events" (LePore 1985, p. 153), by analysing "A shortcircuit caused a fire at 6 p.m. in the bedroom" in the form

$$\exists e \exists f\,(\text{shortcircuit}(e) \wedge \text{fire}(f) \wedge \text{at}(e, 6\,\text{p.m.}) \wedge \text{in}(e, \text{the bedroom}) \wedge \text{caused}(e, f)).$$

But whilst this may point to the existence of processes, the Davidsonian strategy can equally well be applied by analysing the sentence along the lines of

$$\exists e\,(\text{causing of a fire by a shortcircuit}(e) \wedge \text{at}(e, 6\,\text{p.m.}) \wedge \text{in}(e, \text{the bedroom})).$$

The general idea that the same thing might be the subject of different predications doesn't itself eliminate differences of view on the nature of occurrents.[8]

[8] The first predicate, "causing of a fire by a shortcircuit(e)", would be better formulated as "causing of a fire by an appreciable potential difference across a path of low resistance(e)", or "combusting caused by an appreciable potential difference across a path of low resistance$e$".

The construal of processes as causings complements the criticism of the relational analysis of the standard billiard ball case mentioned in Sect. 8.2. The state of the balls prior to collision caused the state of the balls after collision, and the event is surely the collision, involving the process of compression and relaxation of the impacting balls. Anscombe (1993, pp. 92–3) observed that we come by our principal knowledge of causality in learning to use causal concepts, many represented by transitive and other verbs of action like "scrape", "push", "wet", "burn", "knock over", "squeeze", etc., to which we could well add "collide", "gravitate" and "heat". This observation is naturally accommodated in the spirit of the preceding remarks by regarding such verbs as predicates of causings.

8.5 Relating Processes to Quantities, Regions and Times

Having disentangled Davidson's general strategy of redescription from some of his more specific views about occurrents, we can go on to use it as a vehicle for describing some of the general features borne by processes as conceived here. One aspect of this strategy that will guide us here is emphasised by Davidson when he says

> a sentence like "John struck the blow" is about two particulars, John and the blow. ... striking is predicated alike of John and the blow[, and t]his symmetry in the treatment of substances and their changes reflects, I think, an underlying conceptual dependence. Substances owe their special importance in the enterprise of identification to the fact that they survive through time. But the idea of survival is inseparable from the idea of surviving certain sorts of change ... [E]vents often play an essential role in identifying a substance. ... (Davidson 1980, p. 175)

Something can be said about the character of both processes and what Davidson calls substances (i.e. continuants in the terminology used here, where "substance" is reserved for chemical substance) by saying how they are related. The mereological structure of processes can be elaborated by considering the mereological structure of the entities to which processes are related, and the discussion will lead to a deepening of the relational character of processes broached in Sect. 8.2 in opposition to Lombard's non-relational view.

Using letters $e, f, g, h, \ldots$ to range over processes, relational predicates can be used to describe processes as being interactions of certain kinds between continuants. The heating of a gas by a hot water bath, for example, involves a process, f, of heating the gas, ρ, by a quantity of water, π, all of which can be put in the form $H(f, \pi, \rho)$. Once a variable is available to refer to a process, we can go on to say how it is related to time, to material objects, how they are related to space and time when involved in the process, how their parts are related to parts of the process, and so on. A convenient starting point is the fact that a process takes place at (during) a certain time:

$$H(e, \pi, \rho) \wedge TP(e, t),$$

where $TP(e, t)$ is read "e takes place at t" and the variable t is understood as before to range over intervals of time with a mereological structure of parts.

The literal reading of the "takes place" locution suggests that there is a place where a process occurs. But whereas continuants occupy a region for a time in virtue of having a certain volume (although it may vary with temperature and pressure), processes are not ascribed volumes and it is difficult to see what to make of the notion of a place of occurrence. The regularity theory of causation needs some such notion in order to pair off concomitant events with one another without circularity, requiring events (relata of the causal relation) to occupy definite regions in order that they can stand in the contiguity relation. In the paradigm case, it takes the cause (and effect) to be an event constituted by the motion of one billiard ball, and the region occupied by this one continuant during the time at issue is taken to define the location of the event. But the temporal contiguity of the events at issue, which precludes their occurring simultaneously, would seem to be incompatible with the requirement of spatial contiguity. How can the two balls abut if not by occupying abutting regions at the same time? But even if the idea can be made intelligible in the simple case of a two-body billiard ball collision, it is not easily generalised to events involving several continuants, such as three-body collisions, while retaining the force of the contiguity condition.

Other theories of causation, such as the transmission theory of Kistler (1998) and the theories of Salmon and Dowe that he criticises, impose similar tight uniform spatio-temporal constraints on events. This seems to be at variance with how events are ordinarily understood. Referents of expressions such as the "solar eclipse in 1918", "Quisling's betrayal of Norway" or "the marriage" might be said to occur in certain vaguely drawn regions (the solar system, Norway, the church), but there is no commitment to specific boundaries. (Quinton 1979 wonders how far the coronation of Queen Elisabeth II extended into the roof of Westminster Abbey.) On the other hand, if there is nothing more to the notion of event location than the locations of the continuants involved in it, these can be specified directly by adding further conjuncts:

$$H(e, \pi, \rho) \wedge TP(e, t) \wedge Occ(\pi, p, t) \wedge Occ(\rho, q, t),$$

where $Occ(\pi, p, t)$ is read as before (Sect. 3.2) as "π occupies (exactly) the region p at t". It is possible to further specify how regions occupied by the several continuants involved are related. In the case of heating a gas by a large water bath, the regions are mereologically separate:

$$H(e, \pi, \rho) \wedge TP(e, t) \wedge Occ(\pi, p, t) \wedge Occ(\rho, q, t) \wedge p \mid q.$$

The separate regions might abut, but this isn't required for radiant heating, and bodies heat themselves by internal processes such as metabolic reactions and internal friction. Contiguity is not a necessary condition of the occurrence of heating. In the case of the diffusion of one quantity through another, the diffusing quantity occupies part of the region occupied by the matter through which it diffuses:

$$D(e, \pi, \rho) \land TP(e, t) \land Occ(\pi, p, t) \land Occ(\rho, q, t) \land p \subseteq q,$$

where $D(e, \pi, \rho)$ is read "e is a diffusing of π through ρ" and "$\subseteq$" is the mereological part relation. But the spatial relations of quantities involved in processes are typically far more complicated than in these cases; consider, for example, the regions occupied by the quantities combusted in a Bunsen burner flame during the course of some interval.

The argument here, then, is that besides being unclear what spatial region a process might be said to take place in, there is no need for the notion (no need to make the takes place predicate into a three-place relation with a further place for a region of space) because the only definite occupiers of spatial regions are the continuants involved in the process, and which regions they occupy can be clearly specified. Doubts about putative locations of processes always seem to turn on unclarities about the continuants involved. Such cases may arise where descriptions of processes are too vague to precisely delimit the quantities involved and for this reason fail to uniquely specify a particular process, as seems to be the case with two examples mentioned by Johanna Seibt in the following passage:

> This move [of identifying the extension of an entity with the geometric region occupied by that entity] loses its plausibility once we turn away from substances [i.e. continuants] and consider activities, such as *the snowing in Aarhus on February 2nd, 2014* or *the Royal Orchestras playing on the Queen's 70th birthday*, i.e. entities that are spatially uniquely located but whose identity is not defined in terms of their location (for instance, because they are spatially superposable). These activities occupy (at each time t in the temporal interval T) a geometric region R, where R can be understood either as the collection of scattered regions that each are continuous and simply connected (the collection of regions occupied by each snow flake or each musician); alternatively, R can be understood as one multiply connected continuous region (the entire region with spatial holes where at t no snow flake or musician is positioned). (Seibt 2014, p. 176)

Is there a definite spatial location of the snowing? The snowing occurs because of certain conditions of temperature and pressure that drive the water out of the air, forcing it to precipitate as snow. These conditions of temperature and pressure are supported by the ambient air, which is also the source of the water constituting the snow flakes, and so is as much involved in the process as the water of the snow flakes. The quantities involved in the process are therefore not limited to the material of the snowflakes, and jointly occupy a connected region during the course of the snowing. But the description given hardly delimits a definite quantity of air involved in the process, and accordingly fails to delimit a definite region occupied by the continuants involved during the course of the process. Although it may be adequate for certain purposes, for example in directing the immediate course of action of certain agents, a description as vague as this falls short of a description of a definite process. Which processes in the atmosphere around the outskirts of Aarhus are parts of the snowing in Aarhus is unclear and the description fails for this reason to identify a definite process. The criterion of process identity at issue is the mereological one elaborated in the next section, not one based on the idea of spatial location.

The same might be said of the Royal Orchestra's playing on the Queen's 70th birthday. Although the orchestra occupies an unconnected region comprising the sum of the regions occupied by each of the musicians, the playing involves the air, without which there would be no sound. But again, the description goes no way to delimiting a definite quantity of air occupying a region within which the sound waves propagate and outside which they don't. On the other hand, putting aside the question of a spatial location, which is not needed for identification, the description could be regarded as referring to a definite process but without specifying exactly which quantities are involved and for this reason not specifying a region within which the sound waves propagate and outside which they don't. As we will see in Sect. 8.7, processes are often described in ways which fail to mention all the quantities involved.

8.6 The Mereological Structure of Processes

Relating processes to spaces and times in this way raises the question of their mereological structure. Proceeding as before, two mereological postulates,

MP1 $\forall e \forall f (e \subseteq f \wedge f \subseteq e \, .\!\supset \, e = f)$

MP2 $\forall e \forall f (e \circ f \;\equiv\; \sim (e \mid f))$,

stipulating an identity criterion and governing mereological relations are laid down for processes, with part, overlapping and proper part defined as usual. The mereological relations of parthood, overlapping, and so forth, sustained by regions of space, times and quantities of matter impart corresponding mereological relations among processes. To illustrate, note first that the takes place relation $TP(e, t)$ is understood to mean that "e takes place at exactly the time t", and can be written in functional form as $TP(e) = t$. Processes have temporal parts, the mereological structures of times and processes being related by the principle

$$t \subset TP(e) \supset \exists f (f \subset e \wedge TP(f) = t),$$

But parts of processes are not confined to temporal parts; they are also generated by spatial parts of the participating continuants. Thus, if e is a diffusing of π through ρ, then the diffusing of any proper spatial part π' of π through a part of ρ during the same time interval will be a proper part of e but not a temporal part. Similarly, in all the spatial parts of a region throughout which there is a temperature gradient, there will be heatings of the matter occupying those parts. If these spatial parts form a partition (of mutually separate parts jointly summing to the whole region), the magnitude of the total heating will be the sum of the magnitudes of the heatings of the matter in each region of the partition.

Non-temporal parts also arise because the sum of any processes taking place at a given time is a process taking place at that time, so that each of the original processes is a part of the sum and takes place at the same time as the sum. Combining processes to form a more complex process in this way is a standard move in thermodynamic reasoning,[9] the naturalness of which is proof against intuitions to the contrary. Processes are often analysed by viewing them as sums, products or differences of other processes. For example, the motion of a mixture of substances is considered to comprise a motion of the solvent relative to the apparatus, and a motion of each of the other substances relative to the solvent for the purposes of obtaining independent flows to which the Onsager relations apply (Miller 1960, pp. 23, 26). The cross-effects dealt with by irreversible thermodynamics are cases where a single cause, such as a temperature gradient in the Seebeck effect, gives rise to what is naturally described as a process with parts consisting of a heating and a flow of current.

Consecutive sequences of processes are naturally summed to yield an overall process. For example, the route to a highly compressed state of a fluid reached via states above the critical point is an alternative process to the route via condensation (first fn. in Sect. 8.3). A simple Carnot cycle, which is described in textbooks as a sum of four processes in sequence, two isothermal processes separated by two adiabatic processes, is another example. A restriction on such summation is needed, however, given the general principle that

$$\forall e\, \exists t\, TP(e, t),$$

i.e. every process takes place at some time, since mereological sums of intervals are restricted (Sect. 1.3) to exclude times with gaps. In that case, there can be no processes with temporal gaps. To this end, sums of processes are restricted to binary sums (and their binary sums, and so on) by an existence axiom along the lines of:

$$C(TP(e), TP(f)) \supset \exists g\, (\forall h\, (h \mid g \;\equiv .\, h \mid e \,\wedge\, h \mid f) \,\wedge \\ TP(g, TP(e) \cup TP(f))),$$

where two times are connected, abbreviated C, if they abut or overlap (Sect. 1.3), and "$\cup$" is the binary sum operation. A particular consequence of not introducing processes as sums of processes satisfying an arbitrary predicate is that there is no universal process including all the processes that there are. Moreover, a least upper bound of two processes (a process having as parts the two processes in question but no other process containing these two) may exist but their sum may not, and so cannot in general be identified with their sum.

Just as processes may have non-temporal parts, so they may be sums of processes taking place at the same time as well as of consecutive processes. A familiar distinction in organic chemistry is that between concerted and non-concerted

[9] As when an ensemble of processes called Carnot cycles is itself taken to be a Carnot cycle; see. e.g., Duhem (1893a, pp. 300–1).

reactions. Concerted reactions are chemical reactions involving both the breaking and formation of bonds that proceed in a single stage, without the creation of intermediates (which is not to say that a transition state is not involved). Examples are nucleophilic aliphatic substitution reactions proceeding by an S_N2 mechanism, such as the substitution of bromine in methylbromide by reaction with sodium cyanide,

$$CH_3Br + NaCN \rightarrow CH_3CN + NaBr.$$

Pericyclic reactions (involving a transition state with a greater number of closed rings than the total number in the reactants) provide further examples. Diels-Alder reactions are pericyclic reactions, the simplest example of which is the reaction of 1,3-butadiene with ethene to form cyclohexene. The Cope rearrangement of octa 2,6-diene at high temperatures to 3,4-dimethylhexa 1,5-diene is another kind of pericyclic reaction. Although these processes involve parts of the initial reactants successively undergoing the reaction, they are, in a sense, simple processes in so far as the same kind of process is taking place throughout. Non-concerted reactions, by contrast, proceed in several stages.

Unlike concerted reactions, the breaking of bonds in the reactants and formation of bonds in the products takes place at different times in non-concerted reactions, which proceed in several stages with the production of an intermediate. Substitution via an S_N1 mechanism, for example, involves the formation of an intermediate carbocation by the severing of a bond and the subsequent formation of the product by formation of a new bond. The hydrolysis of tertiary butyl bromide, $(CH_3)_3CBr$, in water, for example, proceeds via a first, relatively slow, stage in which the reactant dissociates into a cation and a bromide anion, followed by a rapid stage in which the carbocation is neutralised by reaction with water to form an alcohol and a hydrogen cation:

$$(CH_3)_3CBr \rightarrow (CH_3)_3C^+ + Br^-$$
$$(CH_3)_3C^+ + H_2O \rightarrow (CH_3)_3COH + H^+.$$

Chemical reactions are often complex in this sense, proceeding via a number of elementary processes that can be summed to yield the complex chemical reaction. Chemical kinetics is concerned with the study of how the rates of chemical reactions are affected by various factors with a view to arriving at a reaction mechanism elucidating the various stages by which the reaction proceeds. The overall reaction taking reactants to products is decomposed into a number of elementary reactions involving intermediates—species not appearing amongst the reactants or products, but created and destroyed in the elementary reactions. An apparently simple reaction such as the combustion of hydrogen in oxygen, represented by the overall reaction

$$2H_2 + O_2 \rightarrow H_2O,$$

proceeds, over a limited range of temperature and pressure, by the following elementary steps:

$$\begin{aligned} H_2 + O_2 &\rightarrow 2OH, \\ OH + H_2 &\rightarrow H_2O + H, \\ H + O_2 &\rightarrow OH + O, \\ O + H_2 &\rightarrow OH + H, \\ H &\rightarrow \tfrac{1}{2}H_2 \end{aligned}$$

The overall process, which is the sum of elementary reactions of these kinds, involves the production of hydrogen atoms as intermediates. There are competing processes which, on the one hand, lead to an exponential increase in the hydrogen atom concentration and an ignition, and on the other, reduce the hydrogen atom concentration and tend to terminate the overall process. (Under certain conditions, these terminating reactions can involve ternary collisions between a hydrogen atom, an oxygen molecule and a third body, which might be a hydrogen molecule, an oxygen molecule, a water molecule or molecules of a diluting substance; see Scott 1994, pp. 70–2.) In a closed reactor, changes in temperature and pressure affect the balance of these processes in different ways over a range of these two variables, but ignition leads to an explosion. In a constant flow reactor, where reactants are continually replenished and products removed, a steady flame is supported at appropriate pressures (Scott 1994, pp. 67–73).

The elementary reactions making up an overall reaction illustrate one of several facets to the complexity of processes. Chemical reactions may involve other processes apart from transformations of chemical substances. Turing's suggestion that chemical reactions and diffusion can work together to produce stable stationary patterns of concentrations has been extensively investigated, for example by allowing different reactants to diffuse from different sides of a gel strip. A substance diffusing from a location in a heterogeneous mixture may start to react chemically with the ambient substances in an exothermic reaction which warms the material in the reaction zone, enhancing the local rate of reaction. Such interactions and cross-interactions between continuants leads to a complexity in the idea of which objects might be said to be involved in a given process, suggesting a further difficulty with the idea of defining a process as a change in objects, as discussed in the following section.

8.7 The Relational Character of Processes

The examples of heating and diffusion in Sect. 8.5 illustrate what might be called the relational character of processes (by contrast with Lombard's non-relational view), involving just two quantities of matter. These are simple cases. There is no general restriction on how many bodies might be involved in a process. Only in artificially constructed circumstances, for example, would heating involve just two bodies. A body may be just one among several heating a given object, and except under carefully controlled experimental conditions, any given object would be heating several objects if it heats one. This might suggest that relational predicates like H describing interactions between continuants are multigrade relations. On the other hand, the total heating by a given source might be regarded as the heating of the sum of all the bodies it heats, and this heating might itself be the sum of the heatings by this source of each of the bodies it heats—building, in other words, on the principle

$$H(e, \pi, \rho) \wedge TP(e, t) \wedge H(f, \pi', \rho') \wedge TP(f, t) . \supset . \\ H(e \cup f, \pi \cup \rho, \pi' \cup \rho') \wedge TP(e \cup f, t).$$

Further mereological summation would take into account all the sources heating any of these bodies, and a new mereological sum of these processes is a heating by the sum of the sources of the sum of the sinks.

Even if this approach works for heating it may not be appropriate for processes of other kinds involving many continuants. A multigrade relation seems necessary to accommodate multi-body collisions unless collections in the sense of plural quantification are introduced.

The relational character of processes also finds expression in other ways, which can be illustrated by a consideration of some examples. This will take us a long way from the view of processes as bound up with single continuants in the way suggested by the approach which would define them as changes in specific objects.

Consider the heating of a quantity of water in a beaker by the flame of a Bunsen burner. Here the heating is generated by a chemical reaction consisting in the burning of a gas mixture in air. Natural gas is a mixture of several substances in proportions which vary with the source of the natural gas, but usually comprises mainly methane with small proportions of ethane, propane and higher hydrocarbons, hydrogen and non-combustible substances such as carbon dioxide and helium. The combustion reactions are described by chemical "equations" such as

$$CH_4 + 2O_2 \rightarrow CO_2 + 2H_2O,$$

the single arrow indicating that the reaction goes to completion (as opposed to reaching equilibrium with the products). Different equations are written for the combustion of the different combustible substances. A chemical reaction involving an interaction of n reacting substances,

$$a_1A_1 + \cdots + a_nA_n \rightarrow b_1B_1 + \cdots + b_mB_m,$$

where a_i, b_j are the stoichiometric coefficients representing the combining proportions, is described by an n+m+1-place predicate, $R(e, \pi_1, \ldots, \pi_n, \rho_1, \ldots, \rho_m)$, to the effect that e is an instance of the appropriate kind, where π_i is a quantity of the ith reactant substance kind A_i, and ρ_j is a quantity of the jth product substance kind B_j. Chemical equations are so called because the reactions they describe conform to the Lavoisian principle of the indestructibility of matter, so that $R(e, \pi_1, \ldots, \pi_n, \rho_1, \ldots, \rho_m)$ implies that the mereological sum, $\bigcup_{i=1}^{n} \pi_i$, of the reactants is identical with the mereological sum, $\bigcup_{i=1}^{m} \rho_i$, of the products. In the case in hand, each combustion reaction is described by predicates like "is a combustion of such-and-such a quantity of methane and such-and-such a quantity of oxygen to produce such-and-such a quantity of carbon dioxide and such-and-such a quantity of water". (As we have seen in the previous section, many kinds of reaction have standard names, such as the Diels-Alder reaction, or they may be described as of a general kind, such as a condensation reaction or a pericyclic reaction, which covers many more specific kinds of reaction all sharing some common feature such as water being one of the products or the intermediate having more rings than the reactants. There is an extensive taxonomy covering many interrelations which a formalisation should seek to capture.) The combustion in the Bunsen burner is a mereological sum of several such combustions of the quantities of each of the combustible substances going through the burner during the time in question.

The combustions comprising the burning are each strongly exothermic, which is why they contribute to the heating of the quantity of water. Here, in contrast to the case of heating by a large hot water bath which doesn't change in temperature, there is clearly much changing as quantities of the various kinds of gas are converted from reactants to products of the various chemical reactions. Because different continuants are involved in the combustion and the heating, it might be tempting to say that there are two processes here, the burning and the heating. This is what the idea that processes are defined as particular changings would suggest. Or it might be suggested that the combustion caused the heating. But the combustion is (identical with) the chemical reaction, and the chemical reaction is (i.e. has the property of being) exothermic. The reaction doesn't induce a distinct process involving the "transfer" of energy which takes the form of heating; nor is the heating a part of the chemical reaction as elementary steps are part of the overall reaction. The heating is one of the features of the chemical process which is described by calling it an exothermic process. Ideas inspiring suggestions to the contrary have been criticised earlier, and this example is further grist to the mill. What we have is a situation that can be described along the lines of a simplified conjunction

$$B(e, \pi_1, \ldots, \pi_n, \rho_1, \ldots, \rho_m) \wedge H(e, \bigcup_{i=1}^{n} \pi_i, \sigma),$$

figuring just one kind of chemical process of burning described by an n+m+1-place predicate $B(e, \pi_1, \ldots, \pi_n, \rho_1, \ldots, \rho_m)$ in accordance with the conventions mentioned above. H describes the very same process as a heating of some third quantity, σ, (the water in the beaker) distinct from the material, $\bigcup_{i=1}^{n} \pi_i$, involved in the chemical transformation. Of course, this simple conjunction doesn't capture all that is going on. Much else in addition to the water is heated, and the heating is really the sum of all the combustion processes going on in the flame, itself a complex phenomenon the details of which have not been pursued here. But the simplified formulation emphasises the point, which would still hold for a more complete formulation, that one and the same process is both a chemical reaction involving reactants and products and, in addition, a heating involving bodies completely separate from these which undergo no transformation of substance kind.

The relational character of processes, which stands in opposition to the idea that they involve single objects, thus goes beyond what is captured by using a single many-place predicate to describe them. Chemical reactions can, in general, be harnessed in various ways by the imposition of appropriate constraints to heat or perform work, useful or otherwise. An exothermic reaction in the gas phase, conducted in an apparatus allowing a resulting change of volume by a movable piston to maintain atmospheric pressure, will do work against the atmosphere as well as heating something which might be distinct from the atmosphere. Then we would have a situation described by a conjunction of the kind

$$R(e, \pi_1, \ldots, \pi_n, \rho_1, \ldots, \rho_m) \wedge H(e, \bigcup_{i=1}^{n} \pi_i, \sigma) \wedge W(e, \bigcup_{i=1}^{n} \pi_i, \tau),$$

where $R(e, \pi_1, \ldots, \pi_n, \rho_1, \ldots, \rho_m)$ describes e as a chemical reaction involving the material $\bigcup_{i=1}^{n} \pi_i$, and $W(e, \bigcup_{i=1}^{n} \pi_i, \tau)$ describes e as a working by this material on τ. Some reactions involving electron transfer can be so arranged that the only way the transfer can occur is through a conducting wire in what is called a galvanic cell. These reactions can thus be made to deliver electrical work. The arrangement is reversed in an electrolytic cell, in which electric work by an external electromotive force drives a chemical reaction.

Examples like these suggest that the fact that a process e involves an interaction between π and ρ doesn't exclude the possibility of e also being an interaction, perhaps of a different kind, between π and σ, where σ is distinct, and perhaps even separate, from ρ. Such considerations may involve the environment of what are the objects of primary interest in the description of a process. Chemical reactions conducted in the gas phase are often not straightforwardly reproducible in solution in the liquid phase because of interaction with the solvent. Unimolecular nucleophilic substitution reactions proceeding via an S_N1 mechanism (in which the rate of reaction depends on the concentration of the substrate but not the nucleophile), for example, are sensitive to the nature of the solvent. A case in point is the reaction of tertiary butylchloride with sodium azide, in which the chloride group is replaced by the azide group:

$$(CH_3)_3CCl + NaN_3 \rightarrow (CH_3)_3CN_3 + NaCl.$$

The reaction proceeds in two stages, the slower, rate-determining step being the formation of a t-butyl cation, $(CH_3)_3C^+$, which is highly reactive and quickly neutralised by an azide anion. Solvents with a high dielectric constant favour the reaction by making it easier to separate oppositely charged ions (the force between charged particles depending inversely on the dielectric constant of the medium). Solvents that solvate an ion by restructuring so as to associate with it by for example hydrogen bonding also favour the reaction by stabilising the intermediate cation. Solvents like water are conducive to the reaction on both counts. "In cases like these", as Steve Weininger puts it, "the distinction between reaction and context is rather artificial" (2015, p. 233). The medium in which tertiary butylchloride reacts with sodium azide is just as much an interaction involving the medium as one involving tertiary butylchloride and sodium azide. This is reflected in the thermodynamic principles governing the process, which proceeds on the condition that the overall Gibbs free energy is reduced. This has energy (enthalpy) and entropy contributions the relative weights of which vary with temperature, and the entropy contribution depends in large part on changes within the solvent.

8.8 Modality

There is no evident principle of conservation of processes analogous to the conservation of quantities. But like spaces and times, the fact that they evidently don't gain or lose parts over time has a modal analogue that they can't possibly gain new parts or lose those they have. Accordingly, the mereological essentialism applying to spaces, times and quantities (Sect. 3.6) carries over to processes:

MEss (i) $e \subseteq f \supset \Box(e \subseteq f)$ and (ii) $e \not\subseteq f \supset \Box(e \not\subseteq f)$.

A particular burning of a bunsen burner, for example, which didn't contain a combustion of hexane, couldn't have comprised a burning of hexane, even one that replaces a burning of as many moles of methane (or as many grammes, or of that amount whose combustion generates the same amount of heating in the time of the burning, or of any other comparable amount by whatever criterion). A bunsen-burner burning which differs by having a part which is a combustion of hexane replacing a burning of methane is a different process—one that might possibly have occurred when the actual one did, with the hexane possibly occupying the same place as the methane actually did. This seems to me to be the most reasonable, least arbitrary thing to say even on the supposition that the actual combustion of the methane proceeds independently of the other parts of the overall burning process. In actual fact, the finer details of how one chemical reaction proceeds together with others involving substances occupying the same place as the reactants of the first depends to some extent on those accompanying reactions—according to how much heat they impart to the reactants, to what extent the motions of the one interfere with those of the other, perhaps with a dampening, perhaps with a catalytic effect, etc.—so the import of conceptual replacements of this sort is not to be thought of against the background of a stable remainder. This makes it difficult to understand what process identity would consist in if possible replacement were thought to engender a possible variation rather than a possible variant. With mereological essentialism, on the other hand, the rigidity principles for processes,

R1 $e = f \supset \Box e = f$
R2 $e \neq f \supset \Box e \neq f$,

follow given MP1, the ordinary principles for identity and elementary normal modal logic.

Processes have temporal parts, and it might be thought that a process might, for example, be shorter than it is, possibly losing a temporal part. Whatever plausibility there is in this suggestion seems to derive from taking a certain manner of describing occurrents literally and thinking of processes as continuants. Some philosophers have defended this way of viewing occurrents, speaking of the party moving from the garden into the house, the bush fire encroaching on the town, and so forth. Understood in this way, occurrents might be thought capable of possibly losing a part (being shorter) just as individuals might have a shorter lifetime or lose some of their constitutive matter or even a component. What is possibly lost, on this view, is nevertheless a process in the present sense. The conception seems to be analogous to the way individuals have been construed as being constituted of matter: continuant events are thought of as being constituted of processes at a particular time. There will be some maximal process, the sum of all those processes constituting the continuant event at some time, which constitutes the continuant event for its lifetime. A continuant event might then possibly have endured for a shorter time than it actually did by possibly lacking a final part of its constitutive maximal process—not, be it noted, a part of itself but a part of its constituent process. The party might never have gone on into the chilly evening that enticed the guests to move into the house, the bush fire might have burnt out or been extinguished following a change in wind direction before reaching the town, etc. A criterion of identity in terms of sameness of constitutive processes suggests itself, where, as with the corresponding feature for individuals, the processes constituting a continuant event are not in general a necessary feature of the continuant event, and more specific features must determine possible identity. None of this detracts from the mereological essentialism for processes discussed in the preceding paragraph.

Another distinction that might be mentioned in this context is that between processes and accomplishments or achievements, which distinguish themselves by talking place in exactly the time during which they take place:

$$Acc(e) \equiv \exists t\,(TP(e,t) \wedge \forall t'\,(t' \neq t \supset {\sim}TP(e,t'))).$$

An example might concern a reaction going to completion, say the (specific quantity) of hydrochloric acid neutralised the (specific quantity) of calcium carbonate, or it might concern the establishing of an equilibrium, say the (specific quantity) of hot water came to equilibrium with the (specific quantity) of ice. The ongoing processes involved in these accomplishments are activities of reacting and heating which are distributive and cumulative in the time argument, so that in the case of heating,

$$H(e,\pi,\rho) \wedge TP(e,t) \;.\!\supset\; \forall t' \subseteq t\, \exists f \subseteq e\, (H(f,\pi,\rho) \wedge TP(f,t'))$$

$$H(e,\pi,\rho) \wedge TP(e,t) \wedge H(f,\pi,\rho) \wedge TP(f,t') \wedge Ctt' \;.\!\supset .\; H(e \cup f,\pi,\rho) \wedge TP(e \cup f, t \cup t').$$

The sense of involvement in which an accomplishment involves an activity is mereological parthood. Unlike constitution relationships, which stand between entities of different categories, relations between accomplishments and activities involve processes. Accordingly, the principle of mereological essentialism applies and there is no possibility that an accomplishment should have an activity for part of its duration which it doesn't actually have. Neutralisation of part of the hydrochloric acid with zinc instead of the equivalent proportion of calcium carbonate would be a distinct process, and so a distinct accomplishment. Similarly, an accomplishment involving the same quantities of hydrochloric acid and calcium carbonate but taking place at a lower temperature and therefore taking a longer time would again be a different process. There is no possibility of that particular process taking longer than it did.

Chapter 9
Modal Properties of Quantities

9.1 Introduction

Modal issues have been taken up somewhat sporadically in earlier chapters, without the development of any systematic interpretation. There has been much talk of metaphysical necessity in the contemporary literature, but I have shied away from the talk of essences often associated with this. Hale (2013) proposes an essentialist theory of necessity, seeking the essences or natures of things—essential truths which are held to determine the broader class of necessary truths about these things. Knowledge of such things is held to follow Kripke's scheme, whereby necessary truths follow from the bare truths together with the claim, known a priori by philosophical analysis, that the truth implies the necessary truth. Hale maintains that "while the consequents of conditionals of the form $p \supset \Box p$ do not necessarily and always state facts about essence, they do so when the conditionals themselves are consequences of the right sort of general principles" (Hale 2013, p. 268). Such general principles concern the general kinds to which objects belong, and in particular for substances,

> There is … no mystery about how we know, a priori, the general principle needed for the application of Kripke's inferential model to substance. Our knowledge that gold is a certain element—the element with atomic number 79, and that water is a certain compound—the compound composed of molecules in which two hydrogen atoms covalently bond with an oxygen atom—is of course a posteriori, but we know, courtesy of the very meaning of the term, that if a substance has a certain chemical composition, it necessarily does so. … so if a quantity of matter instantiates a certain substance-kind, it is essential to it that [it] does so. (pp. 280–1)

Quantities were said in Sect. 3.6 to be necessarily ponderable, but as claimed there, what is water is not necessarily, let alone essentially, water. This state of affairs is reviewed in Sect. 9.3 below.

Elements are special kinds of substances. A conception of the elements to which many chemists subscribed for a certain phase in the history of the subject has it

P. Needham, *Macroscopic Metaphysics*, Synthese Library 390,
https://doi.org/10.1007/978-3-319-70999-4_9

that elements endure in their compounds and are indestructible in the isolated state. On this view, it seems plausible that what is a particular element is necessarily that element, perhaps essentially so if like Hale we mean something more specific. Several elements are now known to spontaneously transmute to others in a series of natural radioactive decay processes, whilst others may be artificially induced to do so. Transmutation also occurs as a result of fusion in younger stars and disintegration in older stars. But restricting ourselves to the stable, non-radioactive isotopes and constraining possibilities within a realm of physical conditions in which chemical substances feature (not, e.g., on the surface of a neutron star), the conception at issue would seem to be in accord with the second of the two alternatives described by Hooykaas in the following passage:

> If an alchemist wished to remain faithful to scholastic philosophy and had to interpret the transmutation of copper into copper vitriol, he might choose between two opinions.
>
> 1. Copper is destroyed and a new substance is generated, namely, vitriol, which is as homogeneous as copper itself. Vitriol is not hidden in copper and copper is not hidden in vitriol.
> 2. He could say that the properties of copper disappear but its substantial form, its essence, remains. The change was only accidental, and vitriol is only altered copper. (Hooykaas 1949, p. 72)

Copper is a reddish-brown, glistening metal, which properties are not exhibited by copper vitriol (copper sulphate), a white, crystalline solid turning blue when hydrated by contact with water. Whether elements disappear or merely take on another appearance in their compounds was, as we saw in Chap. 5, debated by the ancients. The second option seems closer to the modern view, which might be thought to offer atomic number as the common essence of elements in isolation and combination. But this is not quite so clear if essence is that which determines that *what* is a particular element is necessarily that element, because a problem arises about identifying one and the same quantity in the compound as that in the isolated state said to bear the essential property. This is discussed in the next section.

Since the discussion of these examples by the proponents of essences is unsatisfactory I see no reason to strive after the distinction they seek to make, and "essentially" and "necessarily" are used interchangeably here when modal claims about water are taken up in Sect. 9.3. Some of these claims prove difficult to capture with the conventional sentential operators for necessity and possibility, leading to a discussion of the use of the notion of possible states to achieve the appropriate expressive power in Sect. 9.4. Incorporating possible processes into this scheme is a natural development, pursued in Sect. 9.5.

9.2 Elements and Compounds

The controversy between the ancients over the issue of whether elements are actually present in their compounds continues to haunt modern views about elements and compounds. The Stoic challenge to the Aristotelian conception of compounds as homeomeric mixts might seem to agree better with modern views in as much as it claims that elements are actually present in compounds. But we saw that this calls for a characterisation of elements which the available texts do not, it seems, provide. Aristotle gave a characterisation of the elements that was, in principle, adequate for his theory of chemical combination, namely by specifying distinguishing features of his four elements as they occur in isolation. Of course, for Aristotle, it goes without saying that the elements occur only in isolation. But proponents of the Stoic view must recognise that the Aristotelian characterisation of the elements only holds good for elements in isolation and is not suited for their purpose, which would allow the elements to occur together in compounds. The proponent of the Stoic idea that elements actually occur in compounds owes us an account of what distinguishes the elements in compounds as well as what distinguishes them in isolation. If these accounts differ, as it seems they must, an account should be forthcoming of what is and isn't the same in the isolated and combined states. The Stoic conception hasn't eliminated all traces of the Aristotelian notion of potential presence.

In a philosophical reflection on the notion of an element, the German physical chemist Friedrich Paneth alluded to much the same issue when he wondered in what sense the elements persist in compounds (Paneth 1931). He distinguished between an element in isolation (what he called a simple substance) and an element in a state of chemical combination (a basic substance). We might see here an allusion to Lavoisier's speaking of the base of oxygen, the base of hydrogen, and so forth, in recognition of the fact, as he saw it, that naturally occurring oxygen gas is a compound of caloric and the element base of oxygen, and that water is a compound of base of hydrogen and base of oxygen (and caloric in liquid and gas states). (Lavoisier thus envisioned these "bases" as part both of what we would regard as the elements in isolation and their compounds, echoing the Stoic view.) Although Lavoisier's view was definitely corrected in the course of the nineteenth century in as much as oxygen gas came to be regarded as an element and not a compound, Paneth recognised that the standard properties of the elements tabulated by chemists were properties exhibited in isolation, and these were not exhibited by compounds, or any parts of compounds, formed from those elements. Oxygen, for example, has characteristic properties such as melting point −218.3 °C and boiling point −182.9 °C, which aren't exhibited by any part of a quantity of water (i.e. solid oxygen would melt if simply heated to −218.3 °C, whereas none of a quantity of solid water at the same initial temperature and pressure would), and similarly for anything that might be thought to count as the hydrogen in water. At normal temperature and pressure, water is a liquid whereas hydrogen and oxygen are gases. Moreover, water at temperatures and pressures when it is a gas, and of the same mass as a mixture of hydrogen and oxygen in the proportion of two moles to one, occupies only two thirds the volume of the mixture. Hydrogen and oxygen are no longer isolated in the mixture, but there clearly remains a difference between the way the elements occupy space in the mixture and in the compound. Paneth certainly draws attention to an important issue, but the question will be What bears the basic substance predicate?

Paneth was instrumental in the adoption by the IUPAC of atomic number instead of atomic weight as defining characteristic of an element in 1923. The discovery of isotopes at the turn of the century threatened the systematisation afforded by the periodic table, and Paneth argued against Fajans that the varying properties of isotopes of the same element were not reasonably counted as chemical(ly significant) differences. Whether it can still be maintained that isotopic variants of the elements are the same substance, especially in the light of more recent advances in precision in chemical kinetics, is another matter (van der Vet 1987; Needham 2008a). But finer distinctions of substance within elements as delimited by atomic number don't affect the present point. Atomic number is envisaged by Kripke (1980) as providing the "microstructural essence" of an element, although what he says about his example, gold, is confined to the isolated element. Does atomic number provide the "nature" that was lacking in the Stoic account and make clear how elements can persist in a compound? That depends on how "persist" is construed. The fact that there are atomic nuclei in water which are either of atomic number one or of atomic number eight (i.e. with either one proton or eight) may be taken to justify the familiar idiom that water contains hydrogen and oxygen, or that there is hydrogen and oxygen in water. But then this latter claim should not be interpreted to mean that a quantity of water can be mereologically partitioned into hydrogen and oxygen, i.e. that there are mutually separate parts, the one comprising all and only hydrogen, the other oxygen, which are jointly exhaustive—their sum is identical with the quantity of water. This mereological claim is not justified by merely speaking of atomic nuclei, which don't exhaust the material in a compound. Electrons must also be taken into account, raising the question of what it is that is supposed to bear the property of having such-and-such an atomic number.

Consider oxygen, with atomic number eight. According to the modern atomic theory in terms of which the periodic table is interpreted, this means that the oxygen nucleus has eight protons. The oxygen nucleus is not an oxygen atom, however, and definitely not oxygen. At the very least, electrons must also be brought into the picture. But how? An isolated oxygen atom is electrically neutral in virtue of possessing eight electrons which are ascribed a structure (in the ground state) denoted by the electronic configuration $1s^2 2s^2 2p^4$. Here the abstraction to a single isolated atom is even greater than that involved in abstracting from mixtures and compounds when considering an element such as oxygen in isolation. But the configuration underlies explanations of oxygen's chemical properties such as its formation of compounds in which oxygen exhibits a valency of two.

There are no particles with the electronic configuration $1s^2 2s^2 2p^4$ in water, which is a covalent molecular compound in the gas state at normal pressure below 500 °C. Water has a considerably more complex microstructure, as we have seen, under other conditions, but it is not necessary to pursue the intricate details in these cases for the purposes of the present discussion. In the molecular state the molecules, as originally described by G. N. Lewis, comprise two nuclei each associated with hydrogen covalently bonded to an oxygen nucleus. Each bond involves a shared pair of electrons, the isolated oxygen atom being considered to contribute one electron to each of these two bonds and an isolated hydrogen atom contributing the other. Distinct bonds are less easily recognised in a more modern molecular orbital account, which construes all the electrons as being involved in the molecular orbitals of the molecule, each with at most two electrons in conformity with the Pauli exclusion principle. On all these accounts, there is simply no question of distinguishing one of the two electrons involved in a covalent bond or relevant molecular orbital as being derived from and belonging to oxygen and the other as deriving from and belonging to hydrogen. (This would be so even on the classical picture envisaged by Lewis. The molecular orbital theory counts the electrons indistinguishable in the quantum mechanical sense.) If oxygen and hydrogen particles can't be disentangled in water in the molecular state, however, then exclusive extensions can't be assigned, alluding to Paneth's terminology, to the "is basic oxygen" and "is basic hydrogen" predicates.

The problem is even more pronounced in aromatic compounds such as benzene and species like the carbonate ion, CO_3^{2-}, with delocalised bonding not confined to particular pairs of nuclei. Classical Lewis covalent bonding together with carbon's valency of four would suggest alternating single and double carbon-carbon bonding in benzene and two single C—O bonds, each with a negative charge on the oxygen, together with one double C—O bond in CO_3^{2-}. But spectroscopic studies show no difference in internuclear distance that would correspond to a difference in single and double bonding in either species. The bonding electrons are delocalised, and associated with the molecule or ion as a whole rather than a particular pair of nuclei.

Theoretical chemists have been far more radical in their criticism of the idea of atoms in molecules after the first attempts to save the idea by Heitler and London in the early days of quantum chemistry. Mulliken (1931) promoted the "molecular" view of the molecular orbital approach according to which the molecule is thought of "as a distinct individual built up of nuclei and electrons", by contrast with "the usual atomic point of view" according to which "the molecule is regarded as composed of atoms or of ions held together by valence bonds" (p. 369). James and Coolidge (1933) thought "the implied persistence of the identity of the individual atoms [in the Heitler and London scheme] quite unjustified, and the resulting wave function is a very poor approximation", and subsequent improvements "by grafting upon it the variation principle do not escape the essential shortcomings of the original method—the implication of atomic individuality, and the omission of an essential coordinate" (p. 825).

Covalent bonding is classically contrasted with ionic bonding, illustrated by common salt. This is a compound of sodium and chlorine that is white, with a crystalline structure at ordinary temperature and pressure with a high melting point and with no trace of the soft, greyish metallic lustre of solid sodium or the green-yellow colour and pungent smell of gaseous chlorine at ordinary temperature and pressure. Sodium, with atomic number eleven, has nuclei with eleven protons. But again the sodium nucleus is not a sodium atom, and definitely not sodium. Isolated sodium atoms have an electronic configuration $[Ne]3s^1$, but there are no particles with this configuration in salt, which is an ionic compound containing sodium cations, i.e. positive ions with one less electron than this configuration attributes. So if sodium atoms and sodium cations are the same kind of thing—

let us call them sodium particles—it is not in virtue of a shared property described by the configuration [Ne]$3s^1$. But not even the isolated solid conforms, comprising as it does a close-packed lattice in which individual "atoms" are combined in a metallic structure where the single valence electron contributes to delocalised bands making for weak binding energies and consequently a soft solid with low melting point. Similarly for the chlorine anion, Cl^-. The electronic configuration describes an ideal condition of a single isolated atom ("infinitely separated" from its closest neighbour), thought to be approximated in the gas state at sufficiently high temperature when the covalently bonded diatomic molecules Na_2 and Cl_2 present at lower temperatures are dissociated. Nevertheless, the electronic configuration is listed as a characteristic property alongside melting and boiling points, implying that what these properties are each characteristic of is a substance which is found in solid and liquid phases as well as the gas phase.

On Lewis's classical picture, the anions and cations of an ionic compound are clearly distinguished as separate entities jointly exhausting the material of the compound. The basic predicates "is basic sodium" and "is basic chlorine" can be understood as "is a collection of sodium cations" and "is a collection of chlorine anions", respectively, applying to quantities that do constitute a mereological partition of what "is sodium chloride" applies to. In reality, however, purely ionic structure is an idealisation that is not realised, and even the most typical ionic structures have some covalent character (e.g. Vanquickenborne 1991, pp. 44–6) with a corresponding entangling of the electrons. But even on Lewis's picture, the partition cannot be construed as a partition into simple sodium and simple chlorine, as a straightforward interpretation of the preservation of sodium and chlorine in sodium chloride would suggest.

Sodium is a simple case in so far as its chemistry is uniformly based on the formation of ionic compounds and a universal valency of one. A transition element such as chromium displays many valencies in its various compounds. Cotton et al. (1999, p. 737) illustrate oxidation states (number of electrons transferred to the more electronegative element in the compound if the bonding were ionic) ranging from −IV to VI. For each oxidation state Z, we need a predicate "is basic chromium Z", which would have different extensions for different Z, each of which would be different from that of "is simple chromium". Again, the idealisation underlying the notion of an oxidation state may suggest that Panethian basic predicates describing the parts of a quantity of a transition metal compound make a mereological partition of the quantity, but in reality the electrons are not so readily disentangled.

Faced with this difficulty in delimiting the elemental components in ionic and covalent compounds, it would seem that the straightforward realisation of the Stoic conception has not proved possible, and something remains of the Aristotelian idea that the original ingredients are not present in a mix.

To emphasise what this means for quantities and their properties, I take it that a straightforward realisation of the Stoic conception would entail that one and the same thing that is isolated sodium at one time might become part of a quantity of sodium chloride at another and yet still be sodium. But even on the simple Lewis model of ionic bonding, Paneth's distinction between simple and basic sodium has to be recognised. Thus, the claim that some sodium might combine with chlorine to form sodium chloride might be formulated:

$$\exists\pi\,(Na(\pi,t)\ \wedge\ \exists\rho\,\exists t'\ \Diamond\ (NaCl(\rho,t')\ \wedge\ \pi\subseteq\rho)),$$

where $Na(\pi,t)$ says that π is sodium. But although the quantity, whether simple sodium or part of some sodium chloride, remains the same, it isn't the case that

$$\exists\pi\,(Na(\pi,t)\ \wedge\ \exists\rho\,\exists t'\ \Diamond\ (NaCl(\rho,t')\ \wedge\ \pi\subseteq\rho\ \wedge\ Na(\pi,t'))),$$

where the *Na* predicate is interpreted uniformly—be it as plain "sodium" or the Paneth notions "simple sodium" or "basic sodium". Nor is this formulation made correct by modifying it so that the first occurrence of *Na* is more precisely specified as Na^S—simple sodium—and the second as Na^B—basic sodium, thus:

$$\exists\pi\,(Na^S(\pi,t)\ \wedge\ \exists\rho\,\exists t'\ \Diamond\ (NaCl(\rho,t')\ \wedge\ \pi\subseteq\rho\ \wedge\ Na^B(\pi,t'))).$$

The reason, as we have seen, is that it is not true on the Lewis model that the same quantity can be both simple sodium and basic sodium. There is no sodium such that *it* might be combined (basic) sodium, and no combined (basic) sodium such that *it* might be isolated (simple) sodium. For the same reason, the uniform interpretation of $Na(\pi,t)$ as $Na^S(\pi,t)\vee Na^B(\pi,t)$ doesn't apply to any single thing throughout a time in the middle of which sodium is converted to sodium chloride. We have simply

$$\exists\pi\,\exists\rho\,(Na(\pi,t)\ \wedge\ Cl(\rho,t)\ \wedge\ \exists t'\ \Diamond\ NaCl(\pi\cup\rho,t')),$$

which captures the potentiality to become (part of) some sodium chloride on the part of some simple sodium as the possibility of being part of some sodium chloride though not as sodium, simple sodium or basic sodium. However, it is not the case that "any and every part" of some sodium chloride might be simple sodium and "any and every part" chlorine, along the lines of Aristotle's claim in *DG* II.7 (see Sect. 5.7).

9.3 Water is H_2O Again

The often-repeated claim that water is necessarily H_2O is usually taken to concern water as an essentially isolated substance (ignoring the presence of dissolved substances in low concentration). Understood in this way, water is necessarily H_2O in the sense that the universal equivalence noted in the first section of Chap. 7 can be strengthened to

$$\Box\,\forall\pi\,\forall t\,(Water(\pi,t)\ \equiv\ H_2O(\pi,t)),$$

with the predicates interpreted as described there. The Barcan formula and its converse for quantity and time variables allows the quantifiers to be equivalently shifted outside the scope of the modal operator.

It seems reasonable to suppose that there is a certain analogy between modality and time. Substance and phase predicates are time-dependent because some quantities possessing them at one time do not do so at another, and it is natural to generalise this by saying that quantities possessing such properties at one time may well not do so at another. So this manner of speaking, which has in fact been adopted already, is to be taken literally. Accordingly, any quantity that happens to be water at some time is not necessarily water at that time (or any other) and, in view of the necessary equivalence of being water and being H_2O, water is not necessarily H_2O in the sense that

$$\forall\pi\ \forall t\,(Water(\pi,t)\ \supset\ \Box H_2O(\pi,t)).$$

Water is essentially H_2O but nothing is essentially water (H_2O).[1]

Pursuing the temporal/modal analogy, we might distinguish the cases in which the time at issue is the same and in which it is entirely separate in making positive statements about what might have been otherwise:

(1) $Water(\pi,t)\ \supset\ \Diamond\!\sim\! Water(\pi,t).$

(2) $Water(\pi,t)\ \wedge\ t\mid t'\ .\!\supset\ \Diamond\!\sim\! Water(\pi,t').$

At first sight it may seem that the first case concerns a more formal sense of what might have been whereas the second concerns what might be brought about by a possible chemical process. But what is now water might not now be water because of a possible chemical process that might have occurred. Different senses of possibility cannot be discerned merely on the basis of such distinctions of the times at issue in the predication.

Problems of expressibility can arise with more complex predicates which we have seen can have greater arity than two and apply to several quantities and several times. The "same substance" relation, for example, is expressed by a four-place predicate applying to two quantities and two time variables, $SameSubst(\pi,t_1,\rho,t_2)$. Thus, just as some quantity that was part of the matter constituting the Atlantic Ocean last week was the same substance as the matter presently constituting the Arctic Ice Sheet but no longer is (because it evaporated and was subsequently chemically transformed in photosynthesis), so a similar fate *might* have befallen another part of the water in the Atlantic last week although it in fact did not. This is not captured by the conjunction

$$SameSubst(\pi,t_1,\rho,t_2)\ \wedge\ \Diamond\!\sim\! SameSubst(\pi,t_2,\rho,t_2),$$

[1] For a more general discussion motivating the rejection of natural kind essentialism, see Leslie (2013) .

which doesn't distinguish between the three cases: π is a different substance at t_2 from what it was at t_1 whilst ρ remains the same, π remains the same whilst ρ changes, and they both change. The problem is to specifically capture the sense of it being possible *for* π to be a different substance at t_2 from ρ at t_2 although it was in fact the same at t_1.

Predicates giving rise to such problems are not to be ignored. Consider

No liquid water could ever be colder than ice ever could be (under normal pressure).

It is difficult to see how this can be expressed in terms of the usual sentential modal operators. It is not captured by, for example,

$$\Box\, \forall \pi\, \forall t_1\, \forall \rho\, \forall t_2\, (\mathit{LiquidWater}(\pi, t_1) \;\wedge\; \Diamond\, \mathit{Ice}(\rho, t_2)\; .\supset {\sim} C(\pi, t_1, \rho, t_2)),$$

where C expresses the colder than relation. A quantity that might be ice at t_2 might not be (at t_2), but the comparison is between a quantity as it would be when liquid water and a quantity as it would be when ice. The claim is not the false one that a quantity cannot be colder at one time than another at another time.

Again,

> Liquid water under pressure might be warmer than liquid water ever could be under normal conditions,

which is not captured by

$$\begin{aligned}&\forall \pi\, \forall t_1\, \forall t_2\, (\mathit{LiquidWaterHighPress}(\pi, t_2) \wedge \\ &\qquad \mathit{LiquidWaterNorPress}(\pi, t_1) \supset \Diamond\, W(\pi, t_2, \pi, t_1)),\end{aligned}$$

where W expresses the warmer than relation. This doesn't capture the modal aspect of how liquid water under normal conditions *ever could be*. Moreover, as it stands the formalised sentence is true simply because a quantity might well be warmer at one time than it is at the same or a different time. But the warmer than relation is said in the present example to stand between a quantity specifically as it would be when it is liquid water under normal pressure and that quantity as it might be when it is liquid water under high pressure. A quantity that is liquid water under normal pressure at t_1 might not be at t_1, but the comparison is with the quantity as it might have been at t_1 when it is liquid water under normal pressure.

Attempting to analyse the same substance, cooler than and warmer than predicates in terms of predicates with lower arity is not particularly plausible. Thermometer readings at different times give a reliable comparison of degrees of warmth only if there is reason to believe that the same reading corresponds to the same degree of warmth. A body is as warm at one time as it is at another if it neither heats nor is heated by anything (maintains a state of thermodynamic equilibrium) during the intervening period. It is possible that thermodynamic equilibrium is not thus maintained but the body is nevertheless as warm at the one time as it is at the other. Ensuring that circumstance, however, calls for a standard of being as warm as at different times. The absolute Kelvin scale might provide such a (highly impractical) standard. But even this requires criteria for when a body heats another, which again falls back on the notion of thermodynamic equilibrium over time. Similar considerations apply to the same substance relation. A sketch of the Aristotelian procedure for introducing substance predicates on the basis of the same substance relation was given in Sect. 5.6, and later developments have not led to the abandonment of this basic idea. But even if cross-temporal comparisons could be reduced to comparisons of numbers or some such, this would not eliminate the modal aspect of comparisons. Some examples of modal comparisons were given at the end of Sect. 3.6 that could be formulated in terms of rigid mereological relations. But the examples just considered call for other resources.

9.4 Possible States

Some philosophers suggest, in the wake of Arthur Prior's creation of tense logic, that introducing time variables imposes an unnatural construal of what appear to be n-place predicates as n+1-place predicates with something not explicit in natural language. But intuitions about what is a "natural" analysis diverge considerably. The operators "it was the case that" and "it will be the case that" introduced to express simple past and future tense inflections are clumsy devices which can hardly be said to provide a natural analysis. The justification of any approach must hang on systematic considerations, which I would submit includes the fact that indexical expressions cannot be iterated (as the expression of tenses by sentential operators would have it), the fact that explicit reference to times and their relation to tense inflections is an integral part of temporal discourse and the fact that we naturally make claims in the form of cross-temporal comparisons. But this dispute cannot be pursued here and I merely note that making reference to time explicit by introducing time variables is standard scientific practice. Purely temporal comparisons, then, are straightforwardly dealt in terms of relational predicates of several times. What should be said about modal comparisons?

Standard modal logic provides for a formal interpretation of sentential modal operators in terms of an extensional model theory in which truth is relativised to possible worlds. Necessity and possibility could be treated directly in terms of universal and existential quantification over possible worlds. But in the absence of an understanding comparable to that of time in everyday and scientific concerns, the natural resistance to ontological commitment to possible worlds speaks in favour of the standard approach in terms of modal sentential operators. Perhaps modal comparisons give us reason to reconsider this position, however.

The notion of a state might be easier to swallow than a possible world as an entity calling for recognition in our ontology. It was natural to use a vocabulary of states when explaining the import of the examples of modal comparisons, and states are a natural part of scientific discourse. Speaking of the possible state of a body seems less extravagant than of the whole universe (or whatever possible things possible worlds are supposed to be). Once states are introduced as objects, several of them might be amalgamated by suitable algebraic operations, but that is a question for later.

Let us suppose, then, that the warmer than relation is properly expressed by a six-place predicate $W(\pi, t_1, s_1, \rho, t_2, s_2)$ read "π at t_1 in state s_1 is warmer than ρ at t_2 in state s_2". The sentence "Liquid water under pressure might be warmer than liquid water ever could be under normal conditions" could then be formalised as

$$\forall\pi\,\forall t_1\,\forall s_1\,\forall t_2\,(\mathit{LiquidWaterNorPress}(\pi, t_1, s_1) \supset \\ \exists s_2\,(\mathit{LiquidWaterHighPress}(\pi, t_2, s_2) \wedge W(\pi, t_2, s_2, \pi, t_1, s_1))),$$

where "$\mathit{LiquidWaterHighPress}(\pi, t_2, s_2)$" can be understood as an abbreviation of "$\mathit{Water}(\pi, t_2, s_2) \wedge \mathit{Liquid}(\pi, t_2, s_2) \wedge \mathit{HighPress}(\pi, t_2, s_2)$", but the warmer than relation cannot be broken down in this way. Introducing a notation more suggestive of the traditional symbolism for modal sentences, this might be reformulated as

$$\Box^1\,\forall\pi\,\forall t_1\,\forall t_2\,(\mathit{LiquidWaterNorPress}({}^1\pi, t_1) \\ \supset \Diamond^2\,(\mathit{LiquidWaterHighPress}({}^2\pi, t_2) \wedge W({}^2\pi, t_2, {}^1\pi, t_1))),$$

where the state variables have been partly suppressed and the quantifiers that bind them replaced with corresponding modal expressions $\Box^1$ and $\Diamond^2$ binding the superscripts attached to quantity variables. More formally, we write "$\Box^i\,\varphi$" where φ results from some well-formed formula $\forall s_i\,\psi$ by removing $\forall s_i$ and replacing every occurrence of a triple π, t_k, s_i, comprising a quantity variable immediately followed by a time variable immediately followed by a state variable, by a pair comprising a quantity variable immediately preceded by the superscript i, identical with the subscript previously on the state variable, immediately followed by the time variable, thus: ${}^i\pi, t_k$. "$\Diamond^i\,\varphi$" is written where φ results from some well-formed formula $\exists s_i\,\psi$ by removing $\exists s_i$ and making the same kind of replacement. The expressions $\Box^i$ and $\Diamond^i$ serve to convey what is necessary or possible *for* a quantity π when binding an index i attached as a superscripted suffix to an occurrence of π in a formula.

The device of indexing quantity variables has been alluded to already (Sect. 5.6) in trying to capture the Aristotelian idea of possible extremal degrees of the primary determinables of warmth and humidity, and in particular of the claim that there might be something which is at some time as warm as anything ever actually is (or could be).

These new modal expressions are not sentential operators but predicate modifiers, distinguished from the familiar operators by the notable feature that they cannot be iterated. An expression of the kind $\Box^1\,\Box^1\,\varphi$ might be syntactically well formed, but the first modal expression is redundant, like the first quantifier in the expression $\forall x\,\forall x\,\varphi$. Indeed, where φ contains no free occurrence of the index "1" (the variable s_1), then it is generally true that $\Box^1\,\varphi \supset \varphi$. This can hardly be interpreted as the analogue of the characteristic axiom for the modal system T because $\Box^1\,\varphi$ doesn't say that φ is necessary where the index doesn't bind. Moreover, where the index doesn't bind it is also generally true that $\varphi \supset \Box^1\,\varphi$, the import of which is certainly not analogous to that of the schema $\varphi \supset \Box\,\varphi$. The latter is generally known as "trivial" because it collapses modal distinctions, whereas the non-binding schema has no such effect and would be appropriately dubbed "harmless".

Expressed in the same abbreviated notation, another problem case considered above, "No liquid water could ever be colder than ice ever could be (under normal pressure)", is captured by

$$\Box^1 \Box^2 \forall\pi \forall t_1 \forall\rho \forall t_2 (LiquidWater(^1\pi, t_1) \wedge Ice(^2\rho, t_2) . \supset \sim C(^1\pi, t_1, ^2\rho, t_2)).$$

In this example and the last the assumption is that the warmth relations hold between quantities at times when they are in definite states, i.e. states of equilibrium. In that case, each part of any of the quantities is as warm as any other of its parts at one of the times in question, and each part of the one quantity is warmer than any part of the other quantity if the first quantity is warmer than the second. The distributive condition has a modal analogue in which the parts sustain the same modal comparison as the quantities themselves.

How are the states and times involved in predications like $W(\pi, t_2, s_2, \pi, t_1, s_1)$ related? A state of a quantity is maintained for some time but is not in general a permanent feature of the body. A quantity comes to be in a given state and is eventually converted into another. It would seem natural to assign states a duration with a function $T(s)$, and to understand the states and times involved in a predication like $W(\pi, t_2, s_2, \pi, t_1, s_1)$ to comply with the condition:

$$W(\pi, t_2, s_2, \pi, t_1, s_1) \supset . \ t_2 \subseteq T(s_2) \wedge t_1 \subseteq T(s_1).$$

Perhaps this should be generalised so that any predication $\varphi(\pi, t_i, s_j)$ of a triple of consecutive variables of the kind π, t_i, s_j should comply with the schema

$$\varphi(\pi, t_i, s_j) \supset t_i \subseteq T(s_j).$$

We will see how this idea can be brought to bear once possible processes are introduced into the development of these modal ideas in the next section.

9.5 Possible Processes

Boyling (1972) presents a formalisation of classical equilibrium thermodynamics in terms of a set of states of a thermodynamic system some of which are adiabatically accessible from others by virtue of possible adiabatic transitions. Taking the liberty of a certain amount of speculative license inspired by Boyling's theory, suppose we can speak of a set of possible states of a quantity of matter over which processes are possible transforming certain states into others. A predicate $Tr(e, s_1, s_2)$ reads "process e makes a transition from state s_1 to state s_2" and each process transforms one state into another. There might be a particular process e effecting such a transition which is, for example, a heating of a quantity π by some quantity ρ,

$$\exists e (H(e, \rho, \pi) \wedge Tr(e, s_1, s_2)),$$

which transforms $\rho \cup \pi$ from state s_1 to state s_2. The existence claim says that such a transition is possible and $\exists e\, Tr(e, s_1, s_2)$ might be construed as saying that s_2 is accessible from s_1. An adiabatic process is one which is not a heating.

Suppose a primitive predicate "*Real*(e)" distinguishes certain processes as real from the merely possible processes, those processes e such that $\sim Real(e)$. Real processes actually take place, and the *TP* predicate introduced in Sect. 8.5 will be taken to specify when possible processes are conceived as taking place as well as when actual ones do (see below). Real processes may be considered to confer reality on states that are transformed by the process into or from another state. Accordingly, a real state can then be defined by

$$Real(s) \equiv \exists e \exists s' (Real(e) \wedge (Tr(e, s, s') \vee Tr(e, s', s))).$$

It is assumed that there are real processes, and since each process transforms some state into another, there are real states.

(An alternative course would have been to take the distinction between real and merely possible states as primitive and define a real process as one effecting a transformation from one real state to another real state. Both states must be real, since a merely possible process might transform a real state to a merely possible one.)

The expression $\Diamond^i \varphi({}^i\pi)$ for φ being possible for π might then be redefined in more elaborate fashion than was done in the previous section by

$$\Diamond^i \varphi({}^i\pi) \equiv \exists e \exists s_i \exists s_j (Real(s_j) \wedge Tr(e, s_j, s_i) \wedge \varphi({}^i\pi)),$$

where $\varphi({}^i\pi)$ is understood as before. The assumption underlying the definition of what it is for a state to be real doesn't preclude a real state being transformed into a possible, perhaps a merely possible, state if the process is not real. As will transpire, however, the simpler and more general original notion $\Diamond^i \varphi({}^i\pi)$ as introduced in the last section has its uses. So rather than redefining it as just suggested, a new notion ${}^i\Diamond^j$ of *relative possibility* is introduced by defining

$${}^i\Diamond^j \varphi({}^j\pi) \equiv \exists e \exists s_j (Tr(e, s_i, s_j) \wedge \varphi({}^j\pi)),$$

where the state variable s_i is free and the relative possibility is not necessarily tied to a real state. But it can be bound by a quantifier introducing a specific condition to that effect. Expressing the idea that π is actually φ by $\Delta^i \varphi({}^i\pi)$, which confines π to a real state and is defined by

$$\Delta^i \varphi({}^i\pi) \equiv \exists s_i (Real(s_i) \wedge \varphi({}^i\pi)),$$

the unbound index in the expression of relative possibility might then be bound by an existential quantifier confining it to a real state in a formula such as $\Delta^i\, {}^i\Diamond^j\, W({}^j\pi, t_1, {}^i\pi, t_2)$. Alternatively, an initial index of relative possibility might not be necessarily real, but bound by another expression of relative possibility in a construction of the kind $\Delta^i\, {}^i\Diamond^j\, {}^j\Diamond^k\, \varphi({}^k\pi)$.

Note that the circumstance expressed by $\Delta^i\, {}^i\Diamond^j\, W({}^j\pi, t_1, {}^i\pi, t_2)$ is a special case of that expressed by $\Delta^k\ \Delta^i\ {}^i\Diamond^j\, W({}^j\pi, t_1, {}^k\pi, t_2)$. Many states may be actual and the expression of actuality doesn't select a unique one.

A corresponding expression of relative necessity defined by

$${}^i\Box^j\,\varphi \quad \equiv \quad \forall e\,\forall s_j\,(Tr(e, s_i, s_j)\ \supset\ \varphi)$$

renders ${}^i\Box^j\,\varphi$ equivalent with $\sim{}^i\Diamond^j\sim\varphi$, and thus ${}^i\Diamond^j\,\varphi$ with $\sim{}^i\Box^j\sim\varphi$. Since ${}^i\Box^j\,\varphi$ doesn't imply that there actually is a real process e which transforms some state s_i into a state s_j, it is not a general principle that ${}^i\Box^j\,\varphi \supset \Delta^i\,\varphi$. Nor does it even hold that $\Delta^i\,{}^i\Box^j\,\varphi \supset \Delta^j\,\varphi$ since processes are not in general cyclic, transforming a state into itself.

Analogues of principles of the modal system K are forthcoming with the tautology $Tr(e, s_i, s_j)\ \supset.\ \varphi \supset \psi\ :\equiv:\ Tr(e, s_i, s_j)\ \supset\ \varphi\ .\supset.\ Tr(e, s_i, s_j)\ \supset\ \psi$, from which it follows that ${}^i\Box^j\,(\varphi\ \supset\ \psi)\ \supset.\ {}^i\Box^j\,\varphi\ \supset\ {}^i\Box^j\,\psi$. This in turn quickly yields ${}^i\Box^j\,(\varphi\ \vee\ \psi)\ \supset.\, {}^i\Diamond^j\,\varphi \vee {}^i\Box^j\,\psi$. Moreover, from $\vdash\ \varphi$ it follows that $\vdash\ {}^i\Box^j\,\varphi$.

Further analogues of principles of the modal system K are readily forthcoming. Since the expression for relative possibility binds a state variable, indicated in the definition above by the index j, this definition can be more generally written as

$${}^i\Diamond^j\,\varphi({}^j\pi_1, \ldots, {}^j\pi_n) \quad \equiv \quad \exists e\,\exists s_j\,(Tr(e, s_i, s_j)\ \wedge\ \varphi({}^j\pi_1, \ldots, {}^j\pi_n)).$$

When $\varphi({}^j\pi_1, \ldots, {}^j\pi_n)$ takes the form of a conjunction, $\varphi({}^j\pi)\ \wedge\ \psi({}^j\rho)$, the formula ${}^i\Diamond^j\,(\varphi({}^j\pi) \wedge \psi({}^j\rho))$ is equivalent with

$$\exists e\,\exists s_j\,(Tr(e, s_i, s_j)\ \wedge\ \varphi({}^j\pi)\ \wedge\ \psi({}^j\rho)),$$

which implies

$$\exists e\,\exists s_j\,(Tr(e, s_i, s_j)\ \wedge\ \varphi({}^j\pi))\ \wedge\ \exists e\,\exists s_j\,(Tr(e, s_i, s_j)\ \wedge\ \psi({}^j\rho)).$$

Accordingly, ${}^i\Diamond^j\,(\varphi({}^j\pi) \wedge \psi({}^j\rho))$ implies ${}^i\Diamond^j\,\varphi({}^j\pi) \wedge {}^i\Diamond^j\ \psi({}^j\rho)$, but not conversely. Similarly, ${}^i\Diamond^j\,(\varphi({}^j\pi) \vee \psi({}^j\rho))$ is equivalent with

$$\exists e\,\exists s_j\,(Tr(e, s_i, s_j)\ \wedge\ (\varphi({}^j\pi)\ \vee\ \psi({}^j\rho))),$$

which is equivalent with

$$\exists e\,\exists s_j\,(Tr(e, s_i, s_j)\ \wedge\ \varphi({}^j\pi))\ \vee\ \exists e\,\exists s_j\,(Tr(e, s_i, s_j)\ \wedge\ \psi({}^j\rho)),$$

and so with ${}^i\Diamond^j\,\varphi({}^j\pi) \vee {}^i\Diamond^j\,\psi({}^j\rho)$. Substitution in and contraposition of these results yields the corresponding results, ${}^i\Box^j\,\varphi({}^j\pi) \vee {}^i\Box^j\,\psi({}^j\rho)\,.\supset\ {}^i\Box^j\,(\varphi({}^j\pi) \vee \psi({}^j\rho))$ and ${}^i\Box^j\,\varphi({}^j\pi) \wedge {}^i\Box^j\,\psi({}^j\rho)\,.\equiv\ {}^i\Box^j\,(\varphi({}^j\pi) \wedge \psi({}^j\rho))$, for relative necessity.

Processes were assigned a time when taking place by a function $TP(e)$ in Sect. 8.5. This idea can reasonably be retained even for the possible processes considered here since these transform states into others which may be merely possible and these have been assigned times when they obtain by a function $T(s)$. A minimal requirement on the relation between these times is then that the transformation predicate $Tr(e, s_i, s_j)$ implies that the states obtain at times that are connected with (overlap or abut) the time when the process takes place:

$$Tr(e, s_i, s_j) \supset . \mathscr{C}\, T(s_i)TP(e) \wedge \mathscr{C}\, TP(e)T(s_j),$$

where $\mathscr{C}$ is the relation of being connected (i.e. overlapping or abutting) amongst times introduced in Sect. 1.3 where it was symbolised as C. (The script style is used here to facilitate reading of the formulas.) Furthermore, the times when the states obtain should be separate. More specifically, $T(s_i) \mid T(s_j)$ would be a consequence of a condition to the effect that it takes time to change one state to another. This might be formulated by: if $Tr(e, s_i, s_j)$ then $\exists e' \, (e' \subseteq e \wedge B_A\, T(s_i)TP(e')T(s_j))$, where B_A is the relation of "abutting betweenness" defined by

$$B_A\, t_1 t_2 t_3 \equiv . \mathscr{B} t_1 t_2 t_3 \wedge \mathscr{A}\, t_1 t_2 \wedge \mathscr{A}\, t_2 t_3,$$

and $\mathscr{B}$ and $\mathscr{A}$ are the relations of betweenness and abutment defined in Sect. 1.3 and written here in script style. But perhaps allowance should also be made for the case where the states are contiguous stages of a continuous process and the one transforms into the other without an intervening time.

This section began with the image of a space of states as envisaged in classical equilibrium thermodynamics over which a relation of adiabatic accessibility is defined. Processes connecting two equilibrium states by paths in the space are idealised "quasistatic", indefinitely slow processes which never carry the system of bodies out of equilibrium. Such processes are useful for calculating the properties of states, but do not trace the course of processes occurring in the world, which proceed at finite rates, carrying the system out of equilibrium and consequently not tracing paths in the space of equilibrium states. A real-world process, which may well be a merely possible process in the sense introduced above, may carry one equilibrium state to another, but shorter parts of it will transform states of affairs themselves not at equilibrium and undergoing change. Suppose such states of affairs were counted as states in the present speculative extension of the equilibrium space. Any proper temporal part of the process could be considered to transform one such dynamic state into another, the states obtaining at abutting times.

With this extension, the special case of a continuous process transforming contiguous states would involve the states obtaining at times that abut and are each part of the time when the process takes place. Adding this alternative to the one already mentioned, we have

$$Tr(e, s_i, s_j) \supset . A_e\, T(s_i)T(s_j) \vee \exists e' \, (e' \subseteq e \wedge B_A\, T(s_i)TP(e')T(s_j)),$$

where the abbreviated expression $A_e\, T(s_i)T(s_j)$ in the first disjunct is defined by

$$A_e\, T(s_i)T(s_j) \equiv . \mathscr{A}\, T(s_i)T(s_j) \wedge T(s_i) \subseteq TP(e) \wedge T(s_j) \subseteq TP(e).$$

This disjunct would come into play with a continuous process such as the objects in a room uniformly undergoing a change in temperature gradient leading to a state in which the objects have the same degree of warmth. The transitional states of some temporal duration are non-equilibrium states undergoing continuous change. The second disjunct would come into play in a longer view of the situation, concerning non-contiguous states arising in the course of the overall transitional process or over a time during which the temperature gradient had evened out.

Finally, the transforms relation is naturally understood to be related to the direction of time, so that where *SD* is the four-place relation of having the same direction defined in Sect. 1.3.1,

If $T(s_1), T(s_2), T(s_3), T(s_4)$ are four pairwise distinct times and $\exists e\, Tr(e, s_1, s_2) \;\wedge$ $\exists e\, Tr(e, s_3, s_4)$, then $SD(T(s_1), T(s_2), T(s_3), T(s_4))$.

The earlier than relation, $<$, is then defined by

$$t_i < t_j \quad \equiv . \;\; t_i \mid t_j \;\wedge\; \exists e\, \exists s_k\, \exists s_l\, (Tr(e, s_k, s_l) \;\wedge\; SD(t_i, t_j, T(s_k), T(s_l)),$$

and we can say that a state that is transformed into another is earlier than that state:

$$Tr(e, s_i, s_j) \;\;\supset\;\; T(s_i) < T(s_j).$$

Taken in conjunction with what was said about the relation between states and times at the end of the last section, the claim $\Delta^{i\,i}\Diamond^j\, W({}^j\pi, t_1, {}^i\pi, t_2)$ that π might be warmer than ρ would imply that t_1 is later than t_2. $\Delta^{i\,i}\Diamond^j\; \Delta^k\;\, W({}^j\pi, t_1, {}^k\pi, t_2)$, on the other hand, would allow that some state earlier than s_k might have been transformed into s_j, allowing in turn that t_1 and t_2 are the same time.

Something of the interplay between temporal and modal aspects of the features of matter that have been of interest here can be raised by first considering the sentence

Some of any water might be an inflammable gas.

This sentence is naturally construed by taking “might be” as making a reference to a later time in the sense of the following formulation:

$$\Delta^i\, \forall t\, \forall \pi\, \exists \rho\, (Water({}^i\pi, t) \;\wedge\; \rho \subseteq \pi \;\wedge\; \exists t'\, (t < t' \;\wedge\; {}^i\Diamond^j\, IG({}^j\rho, t'))),$$

where the predicate $IG({}^j\rho, t')$ stands for “ρ is an inflammable gas at t'”. In the special case

Some of what is water might be an inflammable gas,

the first verb has a present tense reading rendered by introducing a constant “*n*” for now (the indexical aspect of which is understood to require a change in the interpretation of the constant with the speaker’s changing temporal perspective):

$$\Delta^i\, \forall \pi\, \exists \rho\, (Water({}^i\pi, n) \;\wedge\; \rho \subseteq \pi \;\wedge\; \exists t'\, (n < t' \;\wedge\; {}^i\Diamond^j\, IG({}^j\rho, t'))).$$

Compare this with

Some of what is water might have been an inflammable gas,

which can be interpreted in at least two ways. On one account, "might have been" is interpreted as "might have become", when some process might have led to the part of what is in fact water having become an inflammable gas along the lines of:

$$\Delta^i \forall\pi\, \exists\rho\, (Water({}^i\pi, n) \wedge \rho \subseteq \pi \wedge \exists t_1\, \exists t_2\, (t_1 < t_2 < n \wedge {}^{t_1}\Delta^j\, {}^j\Diamond^k IG({}^k\rho, t_2))),$$

where the expression ${}^{t_1}\Delta^j$ fixes the time of the real state s_j as t_1 in accordance with the following elaborated definition of the triangle:

$${}^t\Delta^i\, \varphi \;\equiv\; \exists s_i\, (Real(s_i) \wedge T(s_i) = t \wedge \varphi).$$

This provides for a determination of the time at which a modality arises, as expressed by phrases of the kind "It was/has been/is/will be possible/necessary for ...". In the present example, if the time t_1 of s_j is fixed in this way then the relative possibility entails that t_2 is in fact later than t_1, and it might well be thought that a better interpretation would not necessarily restrict t_2 to being earlier than now. Accordingly, the first account becomes more simply:

$$\Delta^i \forall\pi\, \exists\rho\, (Water({}^i\pi, n) \wedge \rho \subseteq \pi \wedge \exists t_1\, \exists t_2\, (t_1 < n \wedge {}^{t_1}\Delta^j\, {}^j\Diamond^k IG({}^k\rho, t_2))).$$

An alternative account would not construe the part of what is in fact water as necessarily having been transformed into an inflammable gas, but without any such commitment, as simply possibly having been an inflammable gas. For this purpose, the more general expression of possibility introduced in the last section, making no reference to a state from which the possible one is transformed (although not excluding such a state) is appropriate:

$$\Delta^i \forall\pi\, \exists\rho\, (Water({}^i\pi, n) \wedge \rho \subseteq \pi \wedge \exists t_1\, (t_1 < n \wedge \Diamond^j IG({}^j\rho, t_1))).$$

The corresponding expression for necessity, $\Box^i\varphi$, simply means that φ holds for all s_i, and since there are real states, $\Box^i\varphi \supset \Delta^i\varphi$.

Indexical reference is not at issue in more general claims such as the following

Liquid water can never be warmer than steam ever could be at the same pressure,

which is captured by the formula

$$\Delta^i \forall\pi\, \forall\rho\, \forall t\, \forall t' \sim {}^i\Diamond^j\, (LiquidWater({}^j\pi, t) \wedge {}^i\Diamond^k\, (Steam({}^k\rho, t') \wedge SamePress({}^j\pi, t, {}^k\rho, t') \wedge W({}^j\pi, t, {}^k\rho, t'))).$$

Bringing mereological principles into play, when the times of taking place of two processes are connected, the sum of the processes exists and the successive transformation of states is naturally governed by the principle

$$Tr(e, s_1, s_2) \;\wedge\; Tr(e', s_2, s_3) \;\wedge\; \mathscr{C}\,TP(e)TP(e') \;.\!\supset\; Tr(e \cup e', s_1, s_3).$$

This might be expected to effect a reduction in the modal expressions of possibility. Something more specific is required to reduce the expression ${}^i\Diamond^j{}^j\Diamond^k\,\varphi({}^k\pi)$ to ${}^i\Diamond^k\,\varphi({}^k\pi)$, however. The former implies

$$\exists e\,\exists s_j\,\exists e'\,\exists s_k\,(Tr(e, s_i, s_j) \;\wedge\; Tr(e', s_j, s_k) \;\wedge\; \varphi({}^k\pi)).$$

But this doesn't suffice to establish the existence of the sum, $e \cup e'$, of the two processes involved because it doesn't follow that (the times of taking place of) these two processes are temporally connected. The general principle $Tr(e, s_i, s_j) \supset .\; \mathscr{C}\,T(s_i)TP(e) \;\wedge\; \mathscr{C}\,TP(e)T(s_j)$ stated above will not provide the link. What is needed is

$$Tr(e, s_i, s_j) \;\;\supset.\;\; T(s_i) \subseteq TP(e) \;\wedge\; T(s_j) \subseteq TP(e).$$

Then $Tr(e, s_i, s_j) \wedge Tr(e', s_j, s_k)$ would entail that $TP(e)$ and $TP(e')$ overlap in view of a common part $T(s_j)$, establishing the temporal connectedness of e and e' and so the existence of their sum. The required condition is entailed by the first disjunct in the consequent of the more specific principle governing the temporal implications of the transforms relation, introduced in conjunction with a relaxation in the notion of a state. This can't hold in general since a state may well obtain for some time before a process is induced to bring about a change. But it does hold where the states involved are contiguous stages undergoing a continuous process of change.

A notation for the special case of a relative possibility involving a continuously developing process is provided by an expression of the kind ${}^i\,\widehat{\Diamond}\,{}^j\,\varphi({}^j\pi)$ for continuous relative possibility defined by

$$\begin{aligned} {}^i\widehat{\Diamond}{}^j\,\varphi({}^j\pi) \;\;\equiv\;\; & \exists e\,\exists s_j\,(Tr(e, s_i, s_j) \;\wedge\; \mathscr{A}\,T(s_i)T(s_j) \;\wedge \\ & T(s_i) \subseteq TP(e) \;\wedge\; T(s_j) \subseteq TP(e) \;\wedge\; \varphi({}^j\pi)), \end{aligned}$$

or, using the abbreviation $A_e\,T(s_i)T(s_j)$ introduced above,

$${}^i\widehat{\Diamond}{}^j\,\varphi({}^j\pi) \;\;\equiv\;\; \exists e\,\exists s_j\,(Tr(e, s_i, s_j) \;\wedge\; A_e\,T(s_i)T(s_j) \;\wedge\; \varphi({}^j\pi)).$$

Now the expression ${}^i\,\widehat{\Diamond}\,{}^{jj}\Diamond^k\,\varphi({}^k\pi)$ reduces to ${}^i\Diamond^k\,\varphi({}^k\pi)$ (and ${}^i\,\widehat{\Diamond}{}^{jj}\;\widehat{\Diamond}\;{}^k\,\varphi({}^k\pi)$ to ${}^i\,\widehat{\Diamond}\,{}^k\,\varphi({}^k\pi)$). The corresponding continuous relative necessity equivalent to $\sim{}^i\,\widehat{\Diamond}{}^j\sim\varphi({}^j\pi)$, with ${}^i\,\widehat{\Diamond}{}^j\varphi({}^j\pi)$ equivalent to $\sim{}^i\,\widehat{\Box}{}^j\sim\varphi({}^j\pi)$, is defined by

$${}^i\widehat{\Box}{}^j\,\varphi({}^j\pi) \;\;\equiv\;\; \forall e\,\forall s_j\,(Tr(e, s_i, s_j) \;\wedge\; A_e\,T(s_i)T(s_j) \;.\!\supset\; \varphi({}^j\pi)).$$

The corresponding echo of the S4 reduction principle ${}^i\Box^k\,\varphi({}^k\pi) \;\;\supset\;\; {}^i\widehat{\Box}{}^{jj}\Box^k\varphi({}^k\pi)$ follows by substitution and contraposition from that for possibility.

Bringing the mereological principles into play in this way calls for some clarification of what counts as a real process. The natural course to adopt is that a real process is real through and through: there are no merely possible parts of a real process and only real processes sum to generate real processes. The monadic predicate *Real*(e) is, in other words, distributive and cumulative in the sense

$$Real(e) \wedge Real(f) \wedge \mathscr{C}\, TP(e)TP(f) \;.\!\supset Real(e \cup f),$$

where the third conjunct might be directly replaced by $\exists g\, (g = e \cup f)$.

Do states have a mereological structure? Parthood between states would seem to naturally follow parthood between quantities in the corresponding states in the sense that, where φ is a substance or phase property,

$$\varphi({}^{i}\pi, t) \wedge \varphi({}^{j}\rho, t') \wedge \rho \subseteq \pi \wedge t \subseteq t' \;.\!\supset s_j \subseteq s_i.$$

Further, the connectivity condition for the existence of sums of processes laid down by the mereological principles in Sect. 8.6 allows for the summation of parallel processes, i.e. processes taking place at the same time. This might be taken to suggest that there should be a corresponding principle for the summation of states transformed and transforming into others by parallel processes for which

$$\begin{aligned} Tr(e_1, s_i, s_j) \wedge Tr(e_2, s_k, s_l) \wedge TP(e_1) = TP(e_2) \\ .\!\supset\!. \; Tr(e_1 \cup e_2, s_i \cup s_j, s_k \cup s_l). \end{aligned}$$

However, whilst there might be some plausibility in the principle where the states are real, it is not so obviously true, or at any rate lends itself to misleading interpretation, otherwise. With real states, a picture naturally arises of a finite summation of states generating a partial "state of the universe" for a given time, which might even be likened to a partial possible world, namely a partial actual world. Being the sum of actual states, which are all necessarily compatible, the finite sum of real states is a coherent entity which is reasonably regarded as real. But whilst one possible process might transform a state in which a quantity has a certain degree of warmth to one in which it is warmer, another possible process might transform the same state to one in which the quantity is colder. An existential axiom laying down the existence of binary sums of states according as they occur at connected times,

$$\begin{aligned} \mathscr{C}T(s_i)T(s_j) \supset \exists s_k\, (\forall s_l\, (s_l \mid s_k \equiv .\, s_l \mid s_i \wedge s_l \mid s_j) \wedge \\ T(s_k) = T(s_i) \cup T(s_j)), \end{aligned}$$

would generate sums such as that of the warmer and colder states of the quantity just considered. This sum of possible states would not itself be a possible state in anything like the sense of a partial possible world. And it would seem that from the possibility of π becoming warmer than it is and the possibility of π becoming colder than it is,

$$\Delta^i\, ({}^{i}\Diamond^{j}\, W({}^{j}\pi, t', {}^{i}\pi, t) \wedge {}^{i}\Diamond^{k}\, C({}^{k}\pi, t', {}^{i}\pi, t)),$$

the existence of the binary sums would imply, in accordance with the principles given,

$$\Delta^{i\,i}\Diamond^{j\cup k}\,(W(^{j\cup k}\pi, t', {}^{i}\pi, t)\ \wedge\ C(^{j\cup k}\pi, t', {}^{i}\pi, t))$$

to the effect that π might at the same time possibly be both warmer and colder than it actually is (where the index "$j \cup k$" stands for the sum "$s_j \cup s_k$"). Perhaps this unwanted outcome could be avoided with a more restricted principle of summation, including only one state transformed by any given possible process and one resulting state for that process. But the details are not pursued here.

How does the kind of processes involved in a transition of states come into play? Consider how the warmer than relation is related to the process of heating. If the times are the same, a body π is warmer than a body ρ if it would heat it were they in diathermic contact (i.e. not insulated from one another by an adiabatic wall impervious to heat). Suppose a relation $A(\pi, \rho, t)$ of spatial abutment is defined for quantities analogously to the way spatial abutment was defined for individuals in Sect. 2.3. If each of the quantities is connected, this precludes there being any intervening body which could act as an adiabatic (insulating) wall, and so expresses a condition sufficient for being in diathermic contact. (Heating of particular bodies by others might be possible by other means, for example by radiation or an intervening medium facilitating convection.) The heating would transform the state of ρ in virtue of which π is warmer than ρ into some different state. Thus, being warmer than is related to heating along the lines

$$W(^{j}\pi, t, {}^{i}\rho, t)\ \wedge\ A(\pi, \rho, t)\ .\!\supset\ \exists e\,\exists t'\,\exists s_k\,(H(e, \pi, \rho)\ \wedge\ A_e\,T(s_i)T(s_k)$$
$$\wedge\ Tr(e, s_i, s_k)\ \wedge\ W(^{k}\rho, t', {}^{i}\rho, t)\ \wedge\ W(^{j}\pi, t, {}^{k}\rho, t')).$$

This contains the expression $\exists e\,\exists s_k\,(Tr(e, s_i, s_k)\ \wedge A_e\,T(s_i)T(s_k)\ \wedge\ \ldots)$ indicating a continuous possibility (of ρ attaining an intermediate degree of warmth compared with π's and ρ's original states) in accordance with the definition of continuous relative possibility above. The remaining information characterises the possible process as a possible heating. Being warmer than gives rise to a possibility of continuous heating in propitious circumstances.

Where $W(^{j}\pi, t'', {}^{i}\rho, t)$ rather than $W(^{j}\pi, t, {}^{i}\rho, t)$, and the times are different, essentially the same condition holds under the assumption that π is *as warm* (EW) at t as it is at t'':

$$W(^{j}\pi, t'', {}^{i}\rho, t)\ \supset:EW(^{j}\pi, t'', {}^{i}\pi, t)\ \wedge\ A(\pi, \rho, t)\ .\!\supset\ \exists e\,\exists t'\,\exists s_k\,(H(e, \pi, \rho)$$
$$\wedge\,A_e\,T(s_i)T(s_k)\ \wedge\ Tr(e, s_i, s_k)\ \wedge\ W(^{k}\rho, t', {}^{i}\rho, t)\ \wedge\ W(^{j}\pi, t, {}^{k}\rho, t'')).$$

A sufficient condition for a body being as warm at one time as at another is that the body neither heats nor is heated by any body at any time within these times or during the intervening period.

Processes standing in contrast to heatings—adiabatic processes—are rare, calling for special circumstances (contrived by human hand) in which the bodies undergo the processes within perfectly insulated walls. Processes are typically heatings, but usually much more besides. The explosive combination of hydrogen and oxygen to form water, for example, is a process of composition which, like many chemical reactions, is exothermic and might also involve mechanical work, for example in breaking a glass bottle containing the original mixture. Such a process will be one of composition of the hydrogen and oxygen which also involves the heating by the sum of this material of the surroundings and the shattering by this material of the glass of the container. As we saw in the last chapter, these are several descriptions of one and the same process, not proper parts of a process of composition. What do constitute proper parts of this process are its temporal parts and the elementary processes by which the overall process of conversion of the two elements to a single compound transpires, as outlined in Sect. 8.6.

This account of the formation of water is still a description of a process taking place in a laboratory with all the attendant simplifications. More realistic descriptions of processes such as a volcanic eruption, stormy weather, the daily goings on in a human digestive system, and so on, would give a better account of what might lead to states of affairs of the kind actually encountered in the world, but may well be beyond our ken. Nevertheless, we might be confident that it is possible for me to have a headache or for snow on Mont Blanc to start moving and create an avalanche although we can't really specify the kind of process that would be involved. But for the world's matter to assemble in such a way as to create a golden mountain or for Descartes' left leg (or what constitutes it for some period) to be annihilated is to stretch the notion of possibility beyond such limits of plausibility. These are not circumstances realisable in any physical state, but call for possible worlds, accessible from one to another by something more closely resembling a leap of faith than a physical process.

Appendix
Summary of Some Main Principles

Some of the more important principles adopted in the body of the book are summarised here. Page references to definitions are given in the Glossary.

Mereological Axioms for Quantities

With separation, |, as the primitive relation and the usual definitions for part, overlap, etc (see glossary for page references), the mereological axioms are

MQ1 $\forall\pi\,\forall\rho\,(\pi \subseteq \rho \wedge \rho \subseteq \pi \;.\!\supset\; \pi = \rho)$,

MQ2 $\forall\pi\,\forall\rho\,(\pi \circ \rho \;\equiv\; \sim(\pi \mid \rho))$,

MQ3 $\exists\pi\,\varphi \;\supset\; \exists\rho\,\forall\sigma\,(\sigma \mid \rho \;\equiv\; \forall\pi(\varphi \supset \sigma \mid \pi))$.

Uniqueness is easily established and a general sum operation can be defined.

Mereological axioms for other kinds of entities are similar, except where there is no general sum and binary sum and product operations are defined instead. In the case of regions, these axioms are

MS3 $\exists p_3\,\forall p_4\,(p_4 \mid p_3 \;\equiv\!.\; p_4 \mid p_1 \,\wedge\, p_4 \mid p_2)$,

MS4 $p_1 \circ p_2 \;\supset\; \exists p_3\,\forall p_4\,(p_4 \mid p_3 \;\equiv\!.\; p_4 \mid p_1 \,\vee\, p_4 \mid p_2)$.

In the case of times, there is a restriction on the existence of binary sums to connected (abutting or overlapping) intervals, and the axioms for binary sums and products are

MT3 $Ct_1t_2 \;\supset\; \exists t_3\,\forall t_4\,(t_4 \mid t_3 \;\equiv\!.\; t_4 \mid t_1 \wedge t_4 \mid t_2)$,

MT4 $t_1 \circ t_2 \;\supset\; \exists t_3\,\forall t_4\,(t_4 \mid t_3 \;\equiv\!.\; t_4 \mid t_2 \vee t_4 \mid t_2)$.

P. Needham, *Macroscopic Metaphysics*, Synthese Library 390,
https://doi.org/10.1007/978-3-319-70999-4

Abutment

For times,

$$\text{Df.}\, A \quad At_1t_2 \;\equiv.\; t_1 \mid t_2 \wedge \exists t_3\,(t_3 \circ t_1 \wedge t_3 \circ t_2 \wedge \forall t_4\,(t_4 \circ t_3 \;\supset.\; t_4 \circ t_1 \vee t_4 \circ t_2)).$$

Two intervals are *connected* if they either abut or overlap:

$$\text{Df.}\, C \quad Ct_1t_2 \;\equiv.\; At_1t_2 \vee t_1 \circ t_2.$$

For regions of space,

$$\text{Df.}\, A \quad Apq \;\equiv.\; p \mid q \;\wedge\; \exists r\,(Box(r) \;\wedge\; r \circ p \;\wedge\; r \circ q \wedge \forall s\,(s \circ r \;\supset.\; s \circ p \;\vee\; s \circ q)),$$

where "Box" is defined by:

$$\text{Df.}\, Box\, Box(s) \;\equiv\; \exists p\, \exists q\, \exists r\,(p\#q \;\wedge\; q\#r \;\wedge\; p\#r \wedge \forall u\,(u \subseteq s \;\equiv\; u \subseteq p, q, r)).$$

This allows that either of two regions *connected with one another* may be scattered, understanding "connected" as a 2-place predicate analogous to the relation of being connected defined, like the analogous relation on times, as abutting or ovelapping. In particular, every region is connected with itself. The 1-place predicate *being a connected region* (not being scattered) is defined as one that can't be exhaustively partitioned into non-abutting regions:

$$\text{Df.}\, Conn \quad Conn(p) \;\equiv\; \forall q\, \forall r\,(q \mid r \;\wedge\; p = q \cup r \;.\!\supset\; Aqr).$$

Occupancy

The occupies relation is interpreted to mean that the continuant in question occupies exactly the region in question for the time in question (Sects. 2.4 and 2.3):

$$Occ(\pi, p, t) \;\wedge\; p \neq q \;.\!\supset\; \sim\! Occ(\pi, q, t),$$

$$Occ(x, p, t) \;\wedge\; p \neq q \;.\!\supset\; \sim\! Occ(x, q, t).$$

The *principle of accumulation* is expressed by:

$$Occ(\pi, p, t) \;\supset\; \forall q\,(\exists t' \subseteq t\, Occ(\pi, q, t') \supset q \subseteq p),$$

$$Occ(x, p, t) \;\supset\; \forall q\,(\exists t' \subseteq t\, Occ(x, q, t') \supset q \subseteq p).$$

The continuity condition is to the effect that

$$Occ(x, q_1, t_1) \wedge Occ(x, q_2, t_2) \wedge At_1t_2 \,.\!\supset\; Cq_1q_2,$$

The region p occupied for a time by a quantity is a sum of its subregions occupied by the quantity during subintervals in the sense that

$$\begin{aligned} Occ(\pi, p, t) \;\supset\; & \forall t_1 \forall t_2\, (t = t_1 \cup t_2 \supset \\ & \exists q_1 \exists q_2\, (Occ(\pi, q_1, t_1) \wedge Occ(\pi, q_2, t_2) \wedge p = q_1 \cup q_2)). \end{aligned}$$

Similarly for individuals,

$$\begin{aligned} Occ(x, p, t) \;\supset\; & \forall t_1 \forall t_2\, (t = t_1 \cup t_2 \supset \\ & \exists q_1 \exists q_2\, (Occ(x, q_1, t_1) \wedge Occ(x, q_2, t_2) \wedge p = q_1 \cup q_2)). \end{aligned}$$

Sums of regions are related to sums of quantities via the further principle that the region occupied by a sum of quantities is the sum of the regions occupied by these quantities. Capitalising on the functional form of the occupies relation, $Occ(\pi, t) = p$, this principle can be neatly expressed in terms of the binary sum as follows:

$$Occ(\pi \cup \rho, t) = Occ(\pi, t) \cup Occ(\rho, t).$$

Static Abutment

For two quantities:

$$SA(\pi, \rho, t) \;\equiv\; \exists p \exists q \forall t' \subseteq t\, (Occ(\pi, p, t') \wedge Occ(\rho, q, t') \wedge Apq),$$

For two individuals:

$$SA(x, y, t) \;\equiv\; \exists p \exists q \forall t' \subseteq t\, (Occ(x, p, t') \wedge Occ(y, q, t') \wedge Apq).$$

Abutment on the Move

$$A(x, y, t) \;\equiv\; \forall t' \subseteq t\, (\widehat{A}(x, y, t') \vee \widetilde{A}(x, y, t')),$$

where

$$\widehat{A}(x, y, t) \;\equiv\; \forall t' \subseteq t\, \exists p \exists q\, (Occ(x, p, t') \wedge Occ(y, q, t') \wedge Apq)$$

and

$$\widetilde{A}(x, y, t) \;\equiv .\; Occ(x, t) \circ Occ(y, t) \wedge \sim (x \oplus_t y) \wedge$$

$$\forall p, q\, (\exists t', t'' \subset t\, (p = Occ(x, t') \cap Occ(y, t') \wedge$$
$$q = Occ(x, t'') \cap Occ(y, t'') \;\wedge\; t' \subset t'') \;\supset\; p \subset q).$$

Similarly for quantities,

$$A(\pi, y, t) \;\equiv\; \forall t' \subseteq t\, (\widehat{A}(\pi, y, t') \vee \widetilde{A}(\pi, y, t')).$$

Constitution

There is a *principle of accumulation*, analogous to the *principle of accumulation for occupation* formulated above, expressed by:

$$Const(x, \pi, t) \supset \pi \;=\; \Sigma\, \rho\, \exists t' \subseteq t\, Const(x, \rho, t').$$

The criterion of identity is ever having the same constitution, i.e.

(CI) $SameConst(x, y, t) \;\supset\; x = y.$

An individual's lifetime is defined as the maximal period the individual is constituted of something:

$$L(t, x) \;\equiv .\; E(x, t) \wedge {\sim}\exists t'\, (t \subset t' \wedge E(x, t')),$$

where the existence predicate, $E(x, t)$, means being constituted of some quantity for time t. This definition presupposes a *principle of restricted lifetime*, that any individual is constituted of some matter for some time but not for any other times:

$$\exists t\, (\exists \pi\; Const(x, \pi, t) \wedge \forall t'\; (t \mid t' \supset {\sim}\exists \pi\, Const(x, \pi, t'))).$$

When individuals x and y occupy the same place, they are constituted of the same matter, and conversely:

(EP) $\exists p\, (Occ(x, p, t) \wedge Occ(y, p, t)) \;\equiv\; \exists \pi\; (Const(x, \pi, t) \wedge Const(y, \pi, t)).$

It follows from this that the principle of restricted lifetime can be equivalently formulated as

$$\exists t\, (\exists p\, Occ(\pi, p, t) \wedge \forall t'\; (t \mid t' \supset {\sim}\exists p Occ(\pi, p, t'))).$$

This stands in contrast with what holds for quantities, which always occupy somewhere:

$$\exists p\, Occ(\pi, p, t).$$

The equivalence principle naturally yields an equivalent formulation of the identity criterion for individuals in terms of the notion of cooccupancy, defined by

$$Coin(x, y, t) \equiv \forall t' \subseteq t \, \exists p \, (Occ(x, p, t') \wedge Occ(y, p, t')),$$

where the order of the quantifiers is important. The notion of having the same constitution throughout some time is equivalent, in view of the equivalence principle, to that of coinciding throughout this same time, and the above criterion of identity is therefore equivalent to one based on individuals coinciding throughout some time:

$$Coin(x, y, t) \supset x = y.$$

There is no time throughout which distinct individuals coincide: at any time, either they don't exist (don't occupy any region) at that time or they don't occupy the same region throughout that time.

"x is a proper spatial part of y at t", symbolised $x \sqsubset_t y$ and also understood as the relation of being a proper component part, is defined as "for all subintervals t' of t, the region occupied by x at t' is a proper part of the region occupied by y at t'":

$$x \sqsubset_t y \equiv \forall t' \subseteq t \, \exists p \, \exists q \, (Occ(x, p, t') \wedge Occ(y, q, t') \wedge p \subset q).$$

A more general reflexive and transitive relation of component parthood, $x \sqsubseteq_t y$ and read "x is a component (spatial) part y at t", can be defined as "x is either a proper spatial part of or coincident with y at t":

$$x \sqsubseteq_t y \equiv . \, x \sqsubset_t y \vee Coin(x, y, t).$$

The symmetric relation, $\oplus_t$, of spatial overlapping can then be defined by analogy with mereological overlapping as the existence of a common component:

$$x \oplus_t y \equiv \exists z \, (z \sqsubseteq_t x \wedge z \sqsubseteq_t y).$$

Lavoisier's Principle of the Indestructibility of Matter

MEss (i) $\pi \subseteq \rho \supset \Box(\pi \subseteq \rho)$ and (ii) $\pi \not\subseteq \rho \supset \Box(\pi \not\subseteq \rho)$.

This modal essentialism means that quantities comply with what is called the single domain interpretation of quantified modal logic, and satisfy the Barcan formula,

BF $\forall \pi \, \Box \varphi \supset \Box \forall \pi \, \varphi$, equivalently, $\Diamond \exists \pi \, \varphi \supset \exists \pi \, \Diamond \varphi$,

and its converse,

CBF $\Box \forall \pi \, \varphi \supset \forall \pi \, \Box \varphi$, equivalently, $\exists \pi \, \Diamond \varphi \supset \Diamond \exists \pi \, \varphi$.

Likewise, times and regions of space are permanent objects which are never created or destroyed. The Barcan formula and its converse therefore holds for entities of these kinds too, along with the principles of mereological essentialism. But

Kripke's variable domain interpretation is applied to individuals, invalidating the Barcan formula and its converse, and formulated by adapting the principles of free logic:[1]

Q1 $\forall x \varphi \supset . Ey \supset (y/x)\varphi$
Q2 $\varphi \equiv \forall x \varphi$, provided x is not free in φ
Q3 $\forall x (\varphi \supset \psi) \supset . \forall x \varphi \supset \forall x \psi$
Q4 $\forall x Ex$
Q5 $\Diamond Ex$

together with the ridigity principles

R1 $x = y \supset \Box x = y$
R2 $x \neq y \supset \Box x \neq y$.

Distributivity and Cumulativity

The distributive condition for a dyadic predicate $\varphi(\pi, t)$ applying to a quantity and a time is

$$\varphi(\pi, t) \wedge \rho \subseteq \pi \wedge t' \subseteq t . \supset \varphi(\rho, t'),$$

and predicates of higher arity applying to all kinds of entities with a mereological structure follow the same pattern. The cumulative condition for $\varphi(\pi, t)$ is

$$\forall \pi' \subseteq \pi \, \forall t' \subseteq t \, \exists \pi'' \subseteq \pi' \, \exists t'' \subseteq t' \, \varphi(\pi'', t'') \supset \varphi(\pi, t).$$

A dyadic predicate $\varphi(\pi, t)$ is *spatially distributive* iff

$$\varphi(\pi, t) \wedge t' \subseteq t \wedge Occ(\rho, t') \subseteq Occ(\pi, t) \wedge \\ Exh(\pi, t) \wedge Exh(\rho, t') . \supset \varphi(\rho, t').$$

where the predicate, $Exh(\pi, t)$, to the effect that π exhausts all the matter in the region it occupies at t and more briefly read "π is exhaustive at t", is defined by

$$Exh(\pi, t) \equiv \pi = \Sigma\rho \, (Occ(\pi, t) = Occ(\rho, t)).$$

The spatial distributive condition cannot be directly compared with the unrestricted distributivity condition. But under the following assumption of conformity between space, time and matter:

[1]For completeness of the basic system of quantified normal modal logic (Garson 1984), a "mixing" rule must be added in addition to the sentential modal principles of system K, namely
If $\vdash \chi \vee \Box(\psi \supset . Ey \supset (y/x)\varphi)$ then $\vdash \chi \vee \Box(\psi \supset \forall x\varphi)$, provided x is not free in ψ.

$$t' \subseteq t \;\wedge\; Exh(\pi, t) \;\wedge\; Exh(\rho, t') \;.\supset.\; Occ(\rho, t') \subseteq Occ(\pi, t) \equiv \rho \subseteq \pi,$$

the spatial distributivity condition simplifies to:

$$\varphi(\pi, t) \;\wedge\; t' \subseteq t \;\wedge\; \rho \subseteq \pi \;\wedge\; Exh(\pi, t) \;\wedge\; Exh(\rho, t') \;.\supset\; \varphi(\rho, t').$$

A spatio-temporal parthood relation of ρ being a part at a subinterval t' of a quantity π at an interval of time t, written $(\rho, t') \sqsubseteq (\pi, t)$, is naturally defined by

$$(\rho, t') \sqsubseteq (\pi, t) \;\equiv.\; t' \subseteq t \;\wedge\; Exh(\pi, t) \;\wedge\; Exh(\rho, t') \;\wedge\; Occ(\rho, t') \subseteq Occ(\pi, t).$$

By the conformity condition, the right-hand side, and so the ostensively time-dependent spatio-temporal parthood relation, implies $\rho \subseteq \pi$. But this is an implication and not an equivalence, so the time dependence has not dropped out of account. With this relation, the spatial distributivity condition can be more compactly expressed as

$$\varphi(\pi, t) \;\wedge\; (\rho, t') \sqsubseteq (\pi, t) \;.\supset\; \varphi(\rho, t').$$

A spatial cumulative condition formulated analogously to the way the distributive condition was restricted to spatial parts takes the form

$$\begin{aligned}&\forall \pi' \forall t' \,(Occ(\pi', t') \subseteq Occ(\pi, t) \;\wedge\; t' \subseteq t \;\wedge\; Exh(\pi, t) \;\wedge\; Exh(\pi', t') \;.\supset\\ &\exists \pi'' \exists t'' \,(Occ(\pi'', t'') \subseteq Occ(\pi', t') \;\wedge\; t'' \subseteq t' \;\wedge\; Exh(\pi'', t'') \;\wedge\; \varphi(\pi'', t'')))\\ &\qquad\qquad\qquad\qquad\qquad\qquad\qquad\qquad \supset\; \varphi(\pi, t).\end{aligned}$$

Neither the unrestricted cumulative condition nor its spatial-parts variant is stronger than the other. With the notion of spatio-temporal parthood in conjunction with the conformity condition, the restricted cumulative condition can be somewhat more compactly expressed by

$$\begin{aligned}&\forall \pi' \forall t' \,((\pi', t') \sqsubseteq (\pi, t) \supset\\ &\qquad \exists \pi'' \exists t'' \,((\pi'', t'') \sqsubseteq (\pi', t') \;\wedge\; \varphi(\pi'', t''))) \;\supset\; \varphi(\pi, t).\end{aligned}$$

Processes

Two mereological postulates governing identity conditions for and mereological relations between process are

MP1 $\forall e \forall f \,(e \subseteq f \;\wedge\; f \subseteq e \;.\supset\; e = f)$

MP2 $\forall e \forall f \,(e \circ f \;\equiv\; \sim (e \mid f)).$

The mereological structures of times and processes are related by the principle

$$t \subset TP(e) \;\supset\; \exists f \,(f \subset e \;\wedge\; TP(f) = t).$$

A restriction on the summation of processes is needed in view of the general principle that every process takes place at some time, i.e.

$$\forall e\, \exists t\, TP(e, t),$$

To this end, sums of processes are restricted to binary sums (and their binary sums, and so on) by an existence axiom along the lines of:

$$C(TP(e), TP(f)) \supset \exists g\, (\forall h\, (h \mid g \equiv .\, h \mid e \wedge h \mid f) \wedge TP(g, TP(e) \cup TP(f))),$$

where two times are connected, abbreviated C.

Modal Properties of Quantities

That it is possible *for* π to φ is expressed by $\Delta^{i\,i}\Diamond^{j}\, \varphi({}^{j}\pi)$, where the actuality and possibility terms are defined, respectively, by

$$\Delta^{i}\, \varphi({}^{i}\pi) \;\equiv\; \exists s_i\, (Real(s_i) \wedge \varphi({}^{i}\pi)),$$

$${}^{i}\Diamond^{j}\, \varphi({}^{j}\pi) \;\equiv\; \exists e\, \exists s_j\, (Tr(e, s_i, s_j) \wedge \varphi({}^{j}\pi)).$$

A corresponding expression of relative necessity defined by

$${}^{i}\Box^{j}\, \varphi \;\equiv\; \forall e\, \forall s_j\, (Tr(e, s_i, s_j) \supset \varphi)$$

renders ${}^{i}\Box^{j}\, \varphi$ equivalent with $\sim{}^{i}\Diamond^{j}\sim\varphi$, and thus ${}^{i}\Diamond^{j}\, \varphi$ with $\sim{}^{i}\Box^{j}\sim\varphi$.

Determination of the time at which a modality arises, as expressed by phrases of the kind "It was/has been/is/will be possible/necessary for …", is provided for by introducing a temporal modification of the actuality modifier along the following lines:

$${}^{t}\Delta^{i}\, \varphi \;\equiv\; \exists s_i\, (Real(s_i) \wedge T(s_i) = t \wedge \varphi).$$

The earlier than relation, <, is defined by

$$t_i < t_j \;\equiv .\; t_i \mid t_j \wedge \exists e\, \exists s_k\, \exists s_l\, (Tr(e, s_k, s_l) \wedge SD(t_i, t_j, T(s_k), T(s_l))),$$

and we can say that a state that is transformed into another is earlier than that state:

$$Tr(e, s_i, s_j) \;\supset\; T(s_i) < T(s_j).$$

Glossary

A	the abutment relation. 12, 21, 34
SA	the static abutment relation. 35, 38
A_e	the abutting sub-period relation. 193
$\triangle^i$	actually. 191
${}^{t_1}\triangle^j$	actually when. 195
B	the betweenness relation. 14, 19
B_A	the abutting betweenness relation. 193
$\mathscr{B}$	the betweenness relation (alternative notation). 193
Box	the property of being a box. 21
$Coin$	the coincide relation. 58
$\mathbb{C}$	the composes relation. 144
$\overline{\mathbb{C}}$	the proper composition relation. 145
$Conn$	the property of being connected. 21
C	the connected relation. 13, 21
$\mathscr{C}$	the connected relation (alternative notation). 193
$Const$	the constitutes relation. 52
$-$	the difference operation. 11, 13, 148
$-$	the complement operation. 11
E	the interior end part relation. 13
Exh	the exhausts relation. 84
E	the existence relation. 54
$\geq_h$	the at least as humid as relation. 101
$\approx_h$	the as humid as relation. 102

P. Needham, *Macroscopic Metaphysics*, Synthese Library 390,
https://doi.org/10.1007/978-3-319-70999-4

$\approx$	the indistinguishability relation. 143
I	the internal abutment relation. 13
Λ	the least upper bound operation. 10
L	the lifetime relation. 54
$\Box^i$	necessary for. 195
${}^i\Box^j$	necessary (relatively) for. 192
Occ	the occupies relation. 31, 37
$\#$	the orthogonal relation. 20
$\circ$	the overlap relation. 9
$\oplus_t$	the component overlap relation. 61
$\odot$	the micro-overlap relation. 145
$\|$	the parallel relation. 20
$\subseteq$	the part relation. 9
$\subset$	the proper part relation. 9
$\not\subseteq$	the not a part relation. 10
$\sqsubset_t$	the proper component part relation. 60
$\sqsubseteq_t$	the component part relation. 61
$\sqsubseteq$	the spatio-temporal part relation. 85
$\Diamond^i$	possible for. 189
${}^i\Diamond^j$	possible (relatively) for. 191
Π	the general product operation. 10, 148
$\cap$	the binary product operation. 10, 13, 19, 148
$Real$	the property of being real. 191
$SameConst$	the same constitution relation. 53
SD	the same direction relation. 16
$SameSubst$	the same substance relation. 102, 186
$\mid$	the separation relation. 9
$\wr$	the micro-separation relation. 144
S	the property of being a stratum. 19
S_i	the ith substance property. 101
$SingleSubst$	the single substance relation. 102
Σ	the general sum operation. 10, 147
$\cup$	the binary sum operation. 10, 13, 19, 147
TP	the takes place relation. 167
T	the state duration function. 190
Tr	the transition relation. 190
U_q	the universe of quantities. 11
$\approx_w$	the as warm as relation. 102
$\geq_w$	the at least as warm as relation. 101

Bibliography

Almotahari, M. (2013). The identity of a material thing and its matter. *Philosophical Quarterly, 64*, 387–406.
Anscombe, G. E. M. (1993). Causality and determinism. In E. Sosa & M. Tooley (Eds.), *Causation* (pp. 88–104). Oxford: Oxford University Press.
Aristotle. (1984). In J. Barnes (Ed.), *The complete works of Aristotle* (Vol. 1). Princeton: Princeton University Press.
Armstrong, H. E. (1927). Poor common salt!. *Nature, 120*(3022), 478.
Atkins, P. W. (1994). *The second law*. New York: W.H. Freeman.
Ayers, M. (1991). Substance: Prolegomena to a realist theory. *Journal of Philosophy, 88*, 69–90.
Baker, L. R. (1997). Why constitution is not identity. *Journal of Philosophy, 94*, 599–621.
Baker, L. R. (2000). *Persons and bodies: A constitution view*. Cambridge: Cambridge University Press.
Bayly, B. (1992). *Chemical change in deforming materials*. Oxford: Oxford University Press.
Bell, J. L. (2008). *A primer of infinitesimal analysis*. Cambridge: Cambridge University Press.
Bennett, J. (1988). *Events and their names*. Oxford: Oxford University Press.
Bohr, N. (1913). On the constitution of atoms and molecules [Part I]. *Philosophical Magazine, 26*, 1–25.
Bordoni, S. (2012). *Taming complexity: Duhem's third pathway to thermodynamcis*. Urbino: Editrice Montefeltro.
Bostock, D. (1995). Aristotle on the transmutation of the elements in De Generatione et Corruptione 1.1–4. In C. C. W. Taylor (Ed.), *Oxford studies in ancient philosophy* (Vol. XIII). Oxford: Clarendon Press.
Bowley, R., & Mariana S. (1999). *Introductory statistical mechanics*. New York: Oxford University Press.
Boyling, J. B. (1972). An axiomatic approach to classical thermodynamics. *Proceedings of the Royal Society A, 329*, 35–70.
Bréhier, É. (1951). *Chrysippe et l'ancien stoïcisme* (2nd ed.). Paris: Presses Universitaires de France.
Brouzeng, P. (1991). Duhem's contribution to the development of modern thermodynamics. In K. Marinás, L. Ropolyi, & P. Szegedi (Eds.), *Thermodynamics: History and philosophy* (pp. 72–87). London: World Scientific.
Callen, H. B. (1985). *Thermodynamics and an introduction to thermostatistics*. New York: John Wiley.
Callender, C. (1999). Reducing thermodynamics to statistical mechanics: The case of entropy. *Journal of Philosophy, 94*, 348–73.

P. Needham, *Macroscopic Metaphysics*, Synthese Library 390,
https://doi.org/10.1007/978-3-319-70999-4

Carathéodory, C. (1909 [1976]). Untersuchungen über die Grundlagen der Thermodynamik. *Mathematische Annalen, 67*, 355–386. Translated as Investigations into the Foundations of Thermodynamics, by J. Kestin in Kestin (Ed.), *The second law of thermodynamics* (pp. 229–256). Stroudsburg: Dowden, Hutchinson and Ross.

Casati, R., & Achille C. V. (1999). *Parts and places: The structures of spatial representation*. Cambridge: MIT Press.

Castellan, G. W. (1964). *Physical chemistry*. Reading: Addison-Wesley.

Chalmers, A. (2009). *The scientist's atom and the philosopher's stone: How science succeeded and philosophy failed to gain knowledge of atoms*. Dordrecht: Springer.

Chauvier, S. (2017). Individuality and aggregativity. *Philosophy, Theory and Practice in Biology, 9*(11), 1–14.

Clarke, B. L. (1981). A calculus of individuals based on 'Connection'. *Notre Dame Journal of Formal Logic, 22*, 204–218.

Clarke, B. L. (1985). Individuals and points. *Notre Dame Journal of Formal Logic, 26*, 204–218.

Clarke, E. (2010). The problem of biological individuality. *Biological Theory, 5*(4), 312–325.

Cleland, C. (1991). On the individuation of events. *Synthese, 86*, 229–254.

Chappell, V. (1973). Matter. *Journal of Philosophy, 70*, 679–696.

Cooper, J. (2004). A note on aristotle on mixture. In F. de Haas, & J. Mansfeld (Eds.), *Aristotle: On generation and corruption book I, symposium aristotelicum* (pp. 315–26). Oxford: Oxford University Press.

Cotton, F. A., Wilkinson, G., Murillo, C. A., & Bochmann, M. (1999). *Advanced inorganic chemistry* (6th ed.). New York: John Wiley.

Davidson, D. (1980). *Essays on actions and events*. Oxford: Clarendon Press.

Davidson, D. (2004). *Problems of rationality*. Oxford: Clarendon Press.

Denbigh, K. (1981). *The principles of chemical equilibrium* (4th ed.). Cambridge: Cambridge University Press.

Denbigh, K. G., & Denbigh, J. S. (1985). *Entropy in relation to incomplete knowledge*. Cambridge: Cambridge University Press.

Dowe, P. (2009). Causal process theories. In H. Beebee, C. Hitchcock, & P. Menzies (Eds.), *The Oxford handbook of causation*. Oxford: Oxford University Press.

Dretske, F. (1967). Can events move? *Mind, 76*, 479–492.

Duhem, P. (1887). Étude sur les travaux thermodynamiqes de M. J. Willard Gibbs. *Bulletin des Sciences Mathématiques, 11*, 122–148 and 159–176. Translated (with original pagination) in Duhem (2011).

Duhem, P. (1892). Notation atomique et hypothèses atomistiques. *Revue des questions scientifiques, 31*, 391–457. Translated by Paul Needham as Atomic notation and atomistic hypotheses. *Foundations of Chemistry, 2*(2000), 127–180.

Duhem, P. (1893a). Commentaire aux principes de la Thermodynamique. Deuxiéme Partie: Le principe de Sadi Carnot et de R. Clausius. *Journal de Mathématiques Pure et Appliquées, 9*, 293–359. Translated (with original pagination) in Duhem (2011).

Duhem, P. (1893b). Une Nouvelle Théorie du Monde Inorganique. *Revue des Questions scientifiques, 33*, 90–133.

Duhem, P. (1895). Les Théories de la Chaleur. *Revue des Deux Mondes, 129*, 869–901; 130, 380–415, 851–868. Translated as Theories of heat in Duhem (2002).

Duhem, P. (1898). On the general problem of chemical statics. *Journal of Physical Chemistry, 2*, 1–42; 91–115

Duhem, P. (1902). *Le mixte et la combinaison chimique: Essai sur l'évolution d'une idée*. Paris: C. Naud; Reprinted Fayard, Paris, 1985. Translated in Duhem (2002).

Duhem, P. (1910). *Thermodynamique et chimie: Leçons élémentaires* (2nd ed.). Paris: Hermann et fils.

Duhem, P. (1954). *The aim and structure of physical theories*. Translated by Philip Wiener of *La théorie physique: son objet – Sa structure* (Paris, 1914). Princeton: Princeton University Press.

Duhem, P. (2002). *Mixture and chemical combination, and related essays*. Translated and ed. by Paul Needham. Dordrecht: Kluwer.

Duhem, P. (2011). *Commentary on the principles of thermodynamics*. Translated and ed. by Paul Needham. Dordrecht: Springer.

Dupré, J., & O'Malley M. (2009). Varieties of living things: Life at the intersection of lineage and metabolism. *Philosophy and Theory in Biology, 1*, 1–25.

Faye, J. (1997). Is the mark method time dependent? In J. Fay, U. Scheffler, & M. Urchs (Eds.), *Perspectives on time* (pp. 215–236). Dordrecht: Kluwer.

Field, H. H. (1980). *Science without numbers: A defence of nominalism*. Oxford: Blackwell.

Franks, F. (2000). *Water: A matrix of life* (2nd ed.). Cambridge: Royal Society of Chemistry.

Frege, G. (1895). Kritische Beleuchtung einiger Punkter in E. Schröders *Vorlesungen über die Algebra der Logik. Archiv für systematische Philosophie, 1*, 433–456. Translated in P. Geach, & M. Black (1966). *Translations from the philosophical writings of gottlob frege*. Oxford: Blackwell.

French, S. (2014). *The structure of the world: Metaphysics and representation*. Oxford: Oxford University Press.

French, S., & Krause, D. (2006). *Identity in physics: A historical, philosophical and formal analysis*. Oxford: Clarendon Press.

Freudenthal, G. (1995). *Aristotle's theory of material substance: Heat and pneuma, form and soul*. Oxford: Clarendon Press.

Frigg, R. (2008). A field guide to recent work on the foundations of statistical mechanics. In D. Rickles (Ed.), *The Ashgate companion to contemporary philosophy of physics*. London: Ashgate.

Garson, J. W. (1984). Quantification in modal logic. In D. Gabbay, & F. Guenthner (Eds.), *Handbook of philosophical logic* (Vol. II, pp. 249–307). Dordrecht: Reidel.

Gibbs, J. W. ([1876] 1948). On the equilibrium of heterogeneous substances. *Transactions of the Connecticut Academy of Arts and Sciences, 3*, pt. 1 (1876), 108–248; 3, pt. 2 (1878), 343–520. Reprinted in *The collected works of J. Willard Gibbs* (Vol. I). New Haven: Yale University Press.

Gillon, B. S. (1992). Towards a common semantics for English count and mass nouns. *Linguistics and Philosophy, 15*, 597–639.

Hahm, D. E. (1985). The Stoic theory of change. *Southern Journal of Philosophy, 13*(Suppl.), 39–56.

Hale, B. (2013). *Necessary beings: An essay on ontology, modality and the relations between them*. Oxford: Oxford University Press.

Hamblin, C. L. (1969). Starting and stopping. *The Monist, 53*, 410–425.

Hamblin, C. L. (1971). Instants and intervals. *Studium Generale, 24*, 127–134.

Hawley, K. (2002). *How things persist*. Oxford: Oxford University Press.

Hawley, K. (2004). Temporal parts. *The Stanford Encyclopedia of Philosophy* (Fall 2004 Edition), ed. E. N. Zalta. http://plato.stanford.edu/entries/temporal-parts.

Heitler, W., & London F. (1927). Wechselwirkung neutraler Atome und homöopolare Bindung nach der Quantenmechanik. *Zeitschrift für Physik, 44*, 455–472. Translated in Hettema, H. (2000) *Quantum chemistry: Classic scientific papers* (pp. 140–55), Singapore: World Scientific.

Heller, M. (1990). *The ontology of physical objects: Four-dimensional hunks of matter*. Cambridge: Cambridge University Press.

Hilgevoord, J. (2002). Time in quantum mechanics. *American Journal of Physics, 70*, 301–306.

Holden, T. (2004). *The architecture of matter: Galileo to Kant*. Oxford: Clarendon Press.

Hooykaas, R. (1949). The experimental origin of chemical atomic and molecular theory before Boyle. *Chymia, 2*, 65–80.

Huntington, E. V. (1904). Sets of independent postulates for the algebra of logic. *Transactions of the American Mathematical Society, 5*, 288–309.

Huntington, E. V. (1913). A set of postulates for abstract geometry, expressed in terms of the simple relation of inclusion. *Mathematische Annalen, 73*, 522–559.

James, H. M., & Coolidge A. S. (1933). The ground state of the hydrogen molecule. *Journal of Chemical Physics, 1*, 825–834.

Jespersen, O. (1924). *The philosophy of grammar*. London: George, Allen and Unwin.

Johnson, W. E. (1921). *Logic* (Vol. I). Cambridge: Cambridge University Press.
Kim, J. (1976). Events as property exemplifications. In M. Brand, & D. Walton (Eds.), *Action theory* (pp. 159–77). Dordrecht: Reidel.
Kistler, M. (1998). Reducing causality to transmission. *Erkenntnis, 48*, 1–24.
Kondepudi, D., & Prigogine, I. (1998). *Modern thermodynamics: From heat engines to dissipative structures*. London: John Wiley.
Kripke, S. A. (1980). *Naming and necessity*. Oxford: Blackwell.
Lang, S. (1969). *Analysis I*. Reading: Addison-Wesley.
Lavoisier, A. -L. (1965). *Elements of chemistry in a new systematic order, containing all the modern discoveries*. Translated by R. Kerr of *Traité élémentaire de Chimie* (Paris, 1789), reprinted. New York: Dover.
Laycock, H. (2006). *Words without objects: Semantics, ontology, and logic for non-singularity*. Oxford: Clarendon Press.
Leonard, H., & Goodman, N. (1940). The calculus of individuals and its uses. *Journal of Symbolic Logic, 5*, 45–55.
Leonardo da Vinci (1938). *The notebooks of Leonardo da Vinci*, Selected English translation by E. MacCurdy, Reynal and Hitchock, London.
LePore, E. (1985). The semantics of action, event, and singular causal sentences. In E. LePore, & B. McLaughlin (Eds.), *Actions and events: Perspectives on the philosophy of Donald Davidson* (pp. 151–61). Oxford: Blackwell.
Leslie, S. -J. (2013). Essence and natural kinds: When science meets preschooler intuition. In T. Szabo-Gendler, & J. Hawthorne (Eds.), *Oxford studies in epistemology* (Vol. 4, pp. 108–165). Oxford: Oxford University Press.
Lesniewski, S. (1916 [1992]). Foundations of the general theory of sets. I. Translated by D. I. Barnett in S. Lesniewski, *Collected works* (Vol. 1, pp. 129–73). S. J. Surma, J. Srzednicki, D. I. Barnett, & F. V. Rickey (Eds.). Dordrecht: Kluwer.
Lewis, G. N. (1916). The atom and the molecule. *Journal of the American Chemical Society, 38*, 762–785.
Li, T., Donadio, D. & Galli, G. (2013). Ice nucleation at the nanoscale probes no man's land of water. *Nature Communications, 4* (article number 1887).
Lieb, E. H., & Yngvason, J. (1999). The physics and mathematics of the second law of thermodynamics. *Physics Reports, 310*, 1–96. Erratum 314, 669.
Lombard, B. (1986). *Events: A metaphysical study*. London: Routledge and Kegan Paul.
Long, A. A., & Sedley, D. N. (1987). *The hellenistic philosophers* (Vol. 1). Cambridge: Cambridge University Press.
Lowe, E. J. (1998). *The possibility of metaphysics*. Oxford: Clarendon Press.
Lowe, E. J. (2006). *The four-category ontology*. Oxford: Clarendon Press.
Maxwell, J. C. (1867). On the dynamical theory of gases. *Philosophical Transactions of the Royal Society of London, 157*, 49–88.
Maxwell, J. C. (2001). *Theory of heat*. New York: Dover.
McKie, D., & Heathcote, N. H. de V. (1935). *The discovery of specific and latent heats*. London: Edward Arnold. Reprinted by Arno Press, New York, 1975.
Messiah, A. M. L., & Greenberg, O. W. (1964). Symmetrization postulate and its experimental foundation. *Physical Review, 136B*, 248–267.
Miller, D. G. (1960). Thermodynamics of irreversible processes. *Chemical Reviews, 60*, 15–37.
Mulliken, R. S. (1931). Bonding power of electrons and theory of valence. *Chemical Reviews, 9*, 347–388.
Needham, P. (1981). Temporal intervals and temporal order. *Logique et Analyse, 24*, 49–61.
Needham, P. (1985). Would cause. *Acta Philosophica Fennica, 38*, 156–182.
Needham, P. (1996). Substitution: Duhem's explication of a chemical paradigm. *Perspectives on Science, 4*, 408–433.
Needham, P. (2002). Duhem's theory of mixture in the light of the Stoic challenge to the aristotelian conception. *Studies in History and Philosophy of Science, 33*, 685–708.

Needham, P. (2004a). Has Daltonian atomism provided chemistry with any explanations? *Philosophy of Science, 71*, 1038–1047.
Needham, P. (2004b). When did atoms begin to do any explanatory work in chemistry? *International Studies in the Philosophy of Science, 8*, 199–219.
Needham, P. (2008a). Is water a mixture?—Bridging the distinction between physical and chemical properties. *Studies in History and Philosophy of Science, 39*, 66–77.
Needham, P. (2008b). Resisting chemical atomism: Duhem's argument. *Philosophy of Science, 75*, 921–931.
Needham, P. (2009a). An Aristotelian theory of chemical substance. *Logical Analysis and History of Philosophy, 12*, 149–164.
Needham, P. (2009b). Reduction and emergence: A critique of Kim. *Philosophical Studies, 146*, 93–116.
Needham, P. (2010). Nagel's analysis of reduction: Comments in defence as well as critique. *Studies in History and Philosophy of Modern Physics, 41*, 163–170.
Needham, P. (2013). Hydrogen bonding: Homing in on a tricky chemical concept. *Studies in History and Philosophy of Science, 44*, 51–66.
Nicholson, D. (2014). The return of the organism as a fundamental explanatory concept in biology. *Philosophy Compass, 9*, 347–359.
Noonan, H. (1993). Constitution is identity. *Mind, 102*, 133–146.
Paneth, F. A. (1931 [1962]). Über die erkenntnistheoretische Stellung des chemischen Elementbegriffs. *Schriften der Königsberger Gelehrten Gesellschaft, Naturwissenschaftliche Klasse, 8*(Heft 4), 101–125. Translated by Heinz. Post as The epistemological status of the chemical concept of element. *British Journal for the Philosophy of Science, 13*, 1–14 and 144–160.
Parsons, T. (1970). An analysis of mass terms and amount terms. *Foundations of Language, 6*, 363–388.
Penrose, R. (1989). *The Emperor's new mind*. Oxford: Oxford University Press.
Pickel, B. (2010). There is no 'Is' of constitution. *Philosophical Studies, 147*, 193–211.
Poincaré, H. (1913). Les conceptions nouvelles de la matière. In H. Bergson, et al. (Eds.), *Le matérialisme actuel* (pp. 49–67). Paris: Flammarion.
Potter, M. (2004). *Set theory and its philosophy*. Oxford: Clarendon Press.
Prior, A. (1967). *Past, present and future*. Oxford: Clarendon Press.
Psillos, S. (2002). *Causation and explanation*. Chesham: Acumen.
Purcell, E. M., & Pound, R. V. (1951). A nuclear spin system at negative temperature. *Physical Review, 81*(2), 279–280.
Putnam, H. (1975). *Philosophical papers* (Vol. 2). Cambridge: Cambridge University Press.
Quine, W. V. (1960). *Word and object*. Cambridge: MIT Press.
Quinton, A. (1964). Matter and space. *Mind, 73*, 332–352.
Quinton, A. (1979). Objects and events. *Mind, 88*, 197–214.
Randlee, D. A., Cui, Z., Cohn, A. G. (1992). A spatial logic based on regions and connection. In B. Nebel, et al. (Eds.), *Principles of knowledge representation and reasoning, proceedings of the third international conference* (pp. 165–176). Los Altos: Morgan Kaufmann.
Rechel, E. E. (1947). The reversible process in thermodynamics. *Journal of Chemical Education, 24*(6), 298–301.
Rescher, N. (1955). Axioms for the part relation. *Philosophical Studies, 6*, 8–11.
Ricci, J. E. (1951). *The phase rule and heterogeneous equilibrium*. Toronto: Van Nostrand.
Robinson, H. (2004). Substance. *The Stanford Encyclopedia of Philosophy* (Fall 2004 Edition), ed. Edward N. Zalta. http://plato.stanford.edu/entries/substance.
Roeper, P. (1983). Semantics for mass terms with quantifiers. *Noûs, 17*, 251–265.
Ruben, D. -H. (1983). Social parts and wholes. *Mind, 92*, 219–238.
Russell, B. (1972). *Principles of mathematics*. London: George Allen and Unwin.
Saunders, S. (2006). Are quantum particles objects? *Analysis, 66*, 52–63.
Scott, R. L. (1977). Modification of the phase rule for optical enantiomers and other symmetric systems. *Journal of the Chemical Society, Faraday Transactions, 73*(11), 356–360.

Scott, S. K. (1994). *Oscillations, waves and chaos in chemical kinetics*. Oxford: Oxford University Press.

Sedley, D. (1982). The Stoic criterion of identity. *Phronesis, 27*, 255–275.

Seibt, J. (2000). The dynamic constitution of things. In J. Faye, U. Scheffler, & M. Urchs (Eds.), *Facts, things, events* (Poznan studies in the philosophy of the sciences and the humanities, Vol. 76, pp. 241–78). Amsterdam: Rodopi.

Seibt, J. (2014). Non-transitive parthood, leveled mereology, and the representation of emergent parts of processes. *Grazer Philosophische Studien, 91*, 165–190.

Sider, T. (2001). *Four-dimensionalism: An ontology of persistence and time*. Oxford: Oxford University Press.

Simons, P. (1987). *Parts: A study in ontology*. Oxford: Clarendon Press.

Simons, P. (2003). Events. In M. J. Loux, & D. W. Zimmerman (Eds.), *The Oxford handbook of metaphysics* (pp. 357–385). Oxford: Oxford University Press.

Simons, P. (2006). Real wholes, real parts: Mereology without algebra. *Journal of Philosophy, 103*, 597–613.

Sklar, L. (1993). *Physics and chance: Philosophical issues in the foundations of statistical mechanics*. Cambridge: Cambridge University Press.

Smith, B., & Varzi, A. C. (2000). Fiat and bona fide boundaries. *Philosophy and Phenomenological Research, 60*, 401–420.

Stout, R. (1997). Processes. *Philosophy, 72*, 19–27.

Sutcliffe, B. T. (1993). The coupling of nuclear and electronic motions in molecules. *Journal of the Chemical Society, Faraday Transactions, 89*, 2321–2335.

Sutcliffe, B., & Woolley, R. G. (2012). Atoms and molecules in classical chemistry and quantum mechanics. In R. F. Hendry, P. Needham, & A. J. Woody (Eds.), *Handbook of the philosophy of science* (Philosophy of chemistry, Vol. 6, pp. 387–426). Amsterdam: Elsevier.

Tarski, A. (1926) [1983]. Foundations of the geometry of solids. Translated of original 1926 article in *Logic, semantics and metamathematics*. Indianapolis: Hackett Publishing Company.

Thomson, J. J. (1983). Parthood and identity across time. *Journal of Philosophy, 80*, 201–221.

Tisza, L. (1977). *Generalized thermodynamics*. Cambridge: MIT Press.

Todd, R. B. (1976). *Alexander of aphrodisias on Stoic physics*. Leiden: E. J. Brill.

van der Vet, P. (1987). *The aborted takeover of chemistry by physics: A study of the relations between chemistry and physics in the present century*. Doctoral dissertation, University of Amsterdam.

van Inwagen, P. (1981). The doctrine of arbitrary undetached parts. *Pacific Philosophical Quarterly, 62*, 123–137.

van Inwagen, P. (1987). When are objects parts? *Philosophical Perspectives, 1*(Metaphysics), 21–47.

van Inwagen, P. (1994). Composition as identity. *Philosophical Perspectives, 8*(Logic and Language), 207–220.

van Inwagen, P. (2006). Can mereological sums change their parts? *Journal of Philosophy, 103*, 614–630.

Vanquickenborne, L. G. (1991). Quantum chemistry of the hydrogen bond. In P. L. Huyskens, W. A. P. Luck, & Th. Zeegers-Huyskens (Eds.), *Intermolecular forces: An introduction to modern methods and results* (pp. 31–53). Heidelberg: Springer.

Varzi, A. (2013). Boundary. *The Stanford Encyclopedia of Philosophy* (Winter 2013 Edition), E. N. Zalta (Ed.), forthcoming http://plato.stanford.edu/archives/win2013/entries/boundary

Vendler, Z. (1967). Verbs and times. In *Linguistics and philosophy*. Ithica: Cornell University Press.

Wald, F. (1896). Chemistry and its laws. *Journal of Physical Chemistry, 1*, 21–33.

Weatherson, B. (2006). Intrinsic vs. Extrinsic properties. *Stanford Encyclopedia of Philosophy*. http://plato.stanford.edu/entries/intrinsic-extrinsic/

Weisheipl, J. A. (1963). The concept of matter in fourteenth century science. In E. McMullin (Ed.), *The concept of matter in Greek and medieval philosophy*. Notre Dame: University of Notre Dame Press.

Wheeler, J. C. (1980). On Gibbs phase rule for optical enantiomers. *Journal of Chemical Physics, 73*, 5771–5777.
Wiggins, D. (1968). On being in the same place at the same time. *Philosophical Review, 77*, 90–95.
Wiggins, D. (1976). The *De Re* 'Must': A note on the logical form of essentialist claims. In G. Evans, & J. McDowell (Eds.), *Truth and meaning* (pp. 285–312). Oxford: Clarendon Press.
Wiggins, D. (2004). *Sameness and substance renewed.* Cambridge: Cambridge University Press.
Williams, C. J. F. (2000). *Aristotle's De Generatione et Corruptione*. Oxford: Clarendon Press.
Weininger, S. J. (2014). Reactivity and its contexts. In U. Klein, & C. Reinhardt (Eds.), *Objects of chemical inquiry* (pp. 203–236). Science History Publications (div of Watson Publishing International).
Woolley, R. G. (1988). Must a molecule have a shape? *New Scientist, 120*(22 Oct.), 53–57.
Zemansky, M. W., & Dittman, R. H. (1981). *Heat and thermodynamics: An intermediate textbook.* London: McGraw-Hill.
Zernike, J. (1951). Vapour pressures of saturated ammonium bicarbonate solutions. *Recueil des travaux chimiques des Pays-Bas, 70*, 711–719.
Zernike, J. (1954). Three-phase curve of ammonium bicarbonate. *Recueil des travaux chimiques des Pays-Bas, 73*, 95–101.

Index

A
abstract objects, 24
abutment, 12–13, 21, 33–38, 167, 193
  on the move, 34
  static, 34, 38
accumulation condition, 33, 36, 38, 139–140
  (AO), 57
  (PA), 52
activities & accomplishments, 12, 177
air, 36
Alexander of Aphrodisias, 99–100
  exclusion principle, 100, 104
Almotahari, 48
Anaxagoras, 92
Anscombe, 166
Aristotle, 11, 29, 77, 83, 86, 89–111, 184–185
  elements, 91–99, 103
    actual & potential, 97
    first definition (by characteristics), 95
    just four, 95
    second (analytic) definition, 96
    transmutation, 92
  equilibrium, 95
  homogeneity, 109–111
  mixt, 91–94
  overwhelming, 94, 125
  primary determinables, 94–96
    bounded, 95, 105–106, 189
  proportions, 96, 110
  spatial parts, 91
  water, 95
Armstrong, 131
artifacts, 51, 53, 55
atmosphere, 36, 48
atomism, 83, 89, 93
  Daltonian, 119
atoms, 136–137, 182, 183
  atomic number, 182
  in molecules, 183
Ayers, 55

B
Barcan formula, 67, 108
  converse, 67
Bayle, 108
Bennett, 154, 164
Berthollet, 118
betweenness, 14, 19
  abutting, 193
Black, 115, 158
Bohr, 120
Boscovich, 27, 114
Bostock, 97
boundaries, 29–30
  bona fide boundary, 30
  fiat boundary, 30, 32
Bowley & Sánchez, 86
Boyling, 190
Bréhier, 100
Brentano, 29

C
calcium carbonate equilibrium, 123–124, 140–142
Callen, 39

P. Needham, *Macroscopic Metaphysics*, Synthese Library 390,
https://doi.org/10.1007/978-3-319-70999-4

caloric, 115
  free, 116
Casati & Varzi, 30
Castellan, 124
Cauchy sequence, 41
causal priority, 17
causation, 152
  process theories, 153
  regularity theory, 167
  relational view, 153, 164–166
Cavendish, 116
change, 129, 152–157
  Cambridge/real, 153
  Russellian, 153, 161
Chappell, 2
chemical potential, 121
Chrysippus, 99–100
Clarke, 12
Clausius, 162
Cleland, 154–157
cohesion, 114
  hydrogen bonding, 132–134
    cooperative, 132
  intermolecular interactions, 129, 131
  van der Waals forces, 130
coincidence, 56, 58, 86
combination, 118–120, 181
  bonding, 131
    covalent, 182
    delocalised, 183
    ionic, 183
  law of definite proportions, 118
  valency, 129, 182
complete first-order theory, 15
  $\aleph_0$-categoricity, 21
composes, $\mathbb{C}$, 143
constitution, 7, 36, 48–71
  modal argument, 51
  unique, 51–52
continuant events, 177
continuants, 1, 29, 31
  changes in, 152
  involvement in processes, 167, 172–176
continuity condition, 33
cooccupancy, *see* coincidence, 86–87, 90
Cooper, 97–98
count nouns, 75
cover, 32
cumulative condition, 78, 177, 197
  generalised, 81–83
  spatial, 85

D
Davidson, 2, 152, 166
  adverbial modification argument, 164–165
  causal relation, 165
Dedekind, 5
Denbigh, 122
Denbigh & Denbigh, 18
descriptive metaphysics, v
difference
  existential prerequisite of, 11
distant from, 31
distributive condition, 77–78, 95, 97, 101, 107, 130–138, 177, 190, 197
  restrictions, 142
  spatial, 84
Dowe, 153
  conserved quantity theory, 153
Duhem, 3, 43, 86, 90, 117, 123, 124, 126, 128, 155, 159–162

E
endurance, 2
entropy, 18, 39, 119, 162
equilibrium, 101, 129, 138, 161
  calcium carbonate, 123–124, 140–142
  dynamic, 129, 132–134
equivalence principle (EP), 57
essence, 179–180
Euclid's second postulate, 14
Euclidean geometry, 5
exhaustive, 84
existence predicate, 54, 67
extensive variables, 39

F
Fajans, 182
Faye, 17
Fine, 48
formula
  compositional, 128
  structural, 128
four-dimensionalists, 2, 50
free logic, 67
Frege, 5
French, 4
French & Krause, 143
Freudenthal, 114
Frigg, 4
fusion, *see* sum

G
Gibbs, 119
Gillon, 76
gunk, 130

H
Hahm, 100
Hale, 179
Hamblin, 12
Hawley, 2, 51, 66
Heitler & London, 120
Heller, 2
Holden, 27, 31, 75, 106–109, 135
  argument from composition, 113–114
Hooykaas, 180
Huntington, 5

I
impenetrability, 50, 56, 86, 98, 103
  Daltonian atoms, 28
indeterminacy relations, 28, 129
indiscernibility, $\approx$, 143
indistinguishability, 130, 142–150
individuals, 36, 47–74
  coincidence, 58
  common component part, $\oplus_t$, 34, 61
  components, 59–65
  countable things, 48
  cover, 32
  criterion of identity
    coincidence, 58
    same constitution, 53–56
  equivalence principle (EP), 57
  existence predicate, 54
  fixed constitution, 51
  interchange of matter, 53
  lifetime, 54
  modal properties, 66–71
  non-connected, 65
  occupying, 31
infinitesimals, 45
intensive properties, 39, 121, 125
intervals, 11–15, 27, 32, 53, 57, 58, 60
  abutment, 12
  betweenness, 14–15
  connected, 13, 81, 193
  earlier than, 194
  infinite divisibility, 15
  sum of, 13
  unity of time, 14
intrinsic properties, 158

J
James & Coolidge, 183
Jespersen, 75
Johnson, 1

K
Kant, 113
Kim, 154
Kondepudi & Prigogine, 42–43
Kripke, 179
  microstructural essence, 182
  necessity of origin, 69–70
  variable domains, 67–69

L
latent heat, 115
  water, 131–132
Lavoisier, 49, 79, 90, 98, 106
  base of, 116, 181
  caloric's elasticity, 117
  decomposition, 114
  elements, 115–118
  indestructibility of matter, 71, 173
Leonard & Goodman, 9
Leonardo da Vinci, 29, 35
LePore, 165
Leslie, 186
Lesniewski, 5
Lewis, 120, 182
Lombard, 2, 153, 155, 157
Lowe, 2, 24

M
macroscopic, 75, 87, 108, 114, 119, 128
  concept, 44
  matter, 135, 138
    continuous, 27, 130
  ontological claims, 151
many-sorted language, 9
mass predicates, 75–77
  grammatical criteria, 75
  relational, 78
material objects, 24, 25, 38
  abutment, 35
  distant from another, 31
  divisibility, 109
  homogeneous, 120
  individuals & quantities, 36, 48
    modal argument, 51
  mobility, 31
  removed from another, 31

mathematical analysis, 38–43
  smooth function, 39
Maxwell, 87, 129
mechanics, 115
mereology, vi, 6, 7, 23–24, 26, 47, 61, 74
  classical axioms, 9, 24
  criterion of identity, 6, 50, 98
  least upper bound, 10
  mereological essentialism, 67, 109
  mereological partition, 130, 142, 182
  principle of extensionality, 6
  principle of transitivity, 6–7
  quasi-, 144–149
  rigid relations, 71
  spatial parts, 84
  spatio-temporal parthood, 85
  triadic relations, 47
mesoscopic scale, 28
microrealm, 4, 42–43, 87, 129
mixture, *see* Aristotle, mixt
  mechanical, 91
  solution, 126
modal comparison, 106, 188
molecules, 120, 129, 130, 138
  in liquid, 135
  isolated, 135, 150
Mulliken, 183

N
nanoscale, 135
necessity
  essentialist theory of, 179
  for, 189
    when, 195
  metaphysical, 179
  relative, 192
    continuous, 196
nominalism, 39
  Field's, 41
  Goodman's, 5
Noonan, 2

O
occupying, 31–33, 130
  exactly, 33, 37, 38, 44–45
occurrents, 1, 151
ontological commitment, 41, 188
operation, 8, 47, 63, 107
organism, 63, 66

P
Paneth, 90, 181–182
  basic oxygen, 183
  simple/basic substance, 181
Parsons, 76
Pauli exclusion principle, 87, 183
Penrose, 143
perdurance, 2
phase, 35, 50, 118, 138–142
  calcium carbonate, 141
  change, 92, 115, 119, 127, 153
    latent heat, 119
  equilibrium, 138
  mereological interpretation, 139–142
  predicates, dyadic, 79
  rule, 120–126, 134, 137, 140–142
    constraints, 122
Philoponus, 97
phlogiston, 117
planets, 70
Plato, 108
plurality, 65
Poincaré, 49
possible for, 187, 189, 191
  when arising, 195
possible worlds, 188
potential/actual parts doctrines, 106–109
Potter, 5
pressure, 39
Prior, 24, 48, 60
  tense logic, 188
processes, 119, 151–178
  as continuants, 177
  burning, 173
  chemical reactions, 173–176
  diffusing, 168
  endpoints, 156
  entropy, 119
  events, 152
  heating, 119, 158, 166–167, 173
  making a transition, 190
  mereological essentialism, 176
  mereological structure, 169–172
  multigrade relations, 173
  radiation, 155
  real, 161
  real/possible, 191
  relation to space, 167
  relational, 154–155, 172–176
  reversible, 156, 160–161
  rigidity principles, 176
  taking place, *TP*, 169, 191

temporal parts, 177
three-body collision, 154
trajectory, 156
product
existential prerequisite of, 10
Proust, 118
puddles, 55
Putnam, 80, 128

Q
quantities
always somewhere, 57
connected, 107, 135
connected with, 104
cooccupying, 86–87
identity conditions, 9, 49
indestructible, 49
mereological essentialism, 67
mereological structure, 9–11, 75–87
no null quantity, 10, 36, 52, 54
spatial parts, 84
three-dimensional continuants, 47
quantum mechanics, 120
zero-point vibration, 129
Quine, 68, 76–78, 80, 83, 135–137
indispensability argument, 43
Quinton, 86, 167

R
rational numbers, 41
real numbers, 41
order-complete, 41
reduction, 2, 4, 43, 151, 165
regions, 5, 18–23
abutment, 21
betweenness, 19
boundaries, 29–30
box, 21
connected, 21, 33
connected with, 21
orthogonal strata, 20
parallel strata, 20
stratum, 19
Reichenbach's mark method, 16–18
relativity, v
removed from, 31, 107
Rescher, 6
Ricci, 140
ridigity principles, 68
rigid relation, 71
Robinson, 50
Roeper, 81
Ruben, 6

S
Salmon, 152
Schrödinger logic, 143
science, v, 25, 86, 153, 155, 164, 188
limits of error, 3, 43
sea, 36
Seibt, 6, 168
separation closure, 14
ship of Theseus, 53, 54
Sider, 2, 47, 50, 56
Simons, 10, 54, 146, 155
Sklar, 4
Smith & Varzi, 29, 36
space, *see* regions, 78
boundaries, 29–30
immobility, 31
spatial parts, 84
exhaustive, 84
spatio-temporal parthood, 85
state, 121, 158, 164, 188
possible, 190–199
real, 191
statistical fluctuations, 42
Stobaeus, 99
Stoic conception of blend, 84, 86, 97, 99–101, 106, 111, 184
blend, 99
fusion, 99
problem of characterising elements, 100
separation, 97, 99
stoichiometric coefficient, 122, 173
stratum, 19
strong supplementation principle, 10, 146
substance, 26, 49, 127–138
& law of definite proportions, 119
Aristotelian, 101
caloric, 115
characteristic features, 121, 133, 136
distinctions of, 115
element, 126, 179
actually/potentially present, 180–185
isotope, 182
transmutation, 180
independent, 123, 134, 140–141
isomers, 128
enantiomorphs, 124
molecular, 130, 136, 137

substance (*cont.*)
predicates, 77–78
dyadic, 79
time-dependency, 80
same, 78–81, 89, 102, 186
time-dependency, 79
simple, 96
single, 91, 92, 102, 121, 126, 134, 139
thermodynamic criterion of purity, 134
time-dependence, 127
sum
binary, 10
definition of, 10
existence of, 10, 51
existential prerequisite of, 10
fusion, 5
operation, 8, 25
region-quantity relation, 37
restricted, 13
unique, 10, 47
universal, 11
unrestricted, 8, 10
Sutcliffe, 120

T
Tarski, 6
temperature, 4, 40–43, 95, 101, 115, 155–157
Kelvin scale, 121
negative absolute, 39
temporal parts, 1
thermodynamics, 39, 56, 86, 89, 115, 120, 159–162
first law, 119, 159
non-equilibrium, 39–41
second law, 119
Thomson, 72, 127
three-dimensionalists, 1, 50
time, *see* intervals, 74, 78
direction of, 15–18, 194
earlier than, 194
existing in, 24
infinite divisibility of, 15
modal analogy, 176, 186
-modal interplay, 194
temporal comparison, 79, 105, 188
unity of, 14
Tisza, 86
triple point, 121

U
universe of discourse, 8

V
van Inwagen
arbitrary undetached parts, 63–65
Special Composition Question, 25–26
sums can change their parts, 71–74
variance, 120
Varzi, 29
Vasa, 55
Vendler, 12

W
water, 35–36, 49, 69, 77, 79, 92, 116, 118, 127–129, 131–138, 179, 182
"H_2O" predicate, 128
bulk, 135
dissociation of, 125, 133, 134
liquid-vapour equilibrium, 138
necessarily H_2O, 185–186
phase-dependent sense, 128
Weininger, 176
Weisheipl, 97
Wheeler, 125
Whitehead, 2
Wiggins, 86
"is" of constitution, 48
necessity, 68
Williams, 89
Woolley, 120, 150

Z
Zernike, 124

Printed in the United States
By Bookmasters